Develop An iPhone App In Less Than a Day With No Programming Skills Required:

iPhone Development So Easy a Complete Novice Can Figure It Out

By Justin Ascott

Minute Help Press

www.minutehelpguides.com

Table of Contents

Chapter 1: Introduction

The iPhone continues to eat up the mobile phone market, and there's no sign that it will be stopping anytime soon. Frankly speaking, there couldn't be a better time to build an iPhone app for your business. And now it's easier than ever.

Why develop an app?

At the very least, it's a great way to get the word out about your business. Research shows that more people are surfing the web through apps instead of the browser. If you have some useful content or a valuable service, you can reach more customers by offering a free app in the App Store. Or, if you have a unique app idea, you can profit from the sales of your app.

These days, everybody wants a system that will allow them to create a single mobile app for all devices. When smartphones were first introduced, this simply wasn't possible. You had to make one app for the iPhone, another for Android, and more for all of the other devices.

But now, with powerful platforms like PhoneGap, you can create a single Javascript and CSS based mobile app that can be quickly transformed into a native iPhone app. The dream of a universal smartphone standard is slowly becoming a reality. Developers will soon rejoice at the simplicity of it all.

This guide is for business owners and mobile app developers who want to use PhoneGap to transform their web app into a native iPhone app. You'll learn how to build your mobile app with Javascript, CSS, and Sencha Touch. You'll also learn how to use the iPhone Software Development Kit to deploy your app to the app store. Once you're finished reading this, you'll have all the tools you need to start developing your first multi-platform app.

How Much Time Will It Take to Develop My App?

Most business owners overestimate the true cost of developing an iPhone app. It's actually much cheaper than you think. The typical business iPhone app isn't much more than a dressed up advertisement. Yes, you'll still have some amount of functionality you've got to program in, but not so much as to distract your user from your brand message. Simple apps like this don't take that much time to develop.

Of course, you can still make your app as feature-filled as you want. Who are we to stop you? If you want to create a game concept that nobody has thought of before, you're more than welcome. We're simply trying to tell you that the barrier to entry is lower than you think, and you'll have plenty of tools and resources to get the job done.

You can expect to spend at least one month on your app. Most of that time won't be spent actually building your app. It will be spent in testing, troubleshooting, and waiting for approval from Apple. Even the simplest apps require a lot of testing. If you want to go for an even higher quality, practically bullet-proof app, you're looking at a two month timeframe. Consider this before you start your project.

O.K. So what will it cost?

Your development costs largely depend on how much equipment you already own. There's a good chance that you own at least one Apple laptop or desktop machine. If so, you will be using it as your development environment. If you don't already own a Mac of some kind, there are ways to turn the Mac Mini into a powerful development environment for a relatively low cost.

I suggest this resource:
http://www.popsci.com/diy/article/2009-02/how-make-iphone-app

Here's a quick rundown of your base development cost:
One Mac with at least 250 GB of disk space and 4GB RAM. Roughly $900
iPhone Developer Program subscription. $99 plus tax.
One iPod Touch. $229 Plus tax.

Other than those costs, that's really all you're looking at. The next section in this guide will show you how to build your web app, which will then be used as the template for your iPhone App. Once you've built your web app, the creative part will be finished. You'll only need to port your app to the iPhone with PhoneGap, and test it out. You'll be well on your way to getting your app on every device in the market.

Chapter 2: How To Turn a Website Into a Mobile Website

It all starts with your mobile website. Even if you don't plan on using PhoneGap to create an iPhone app for your business, you should at least know what you need to do differently when catering to an audience that's accessing your site from different mobile devices. In this section, you'll learn how to adapt your existing website for mobile phones and tablets. If you don't have a website yet, don't worry. We'll show you a quick and easy way to build one from scratch.

How does a mobile website differ from a standard website?

In other words, why can't you just keep your website the way it is? Why do you need to adapt it to smartphones and tablets? What's the big deal anyway? Well here's some insight.

Yes, you could just leave your website untouched, and anyone with a smartphone could still access it. Unfortunately, their experience is going suffer *big time*. All of your content will appear squished and unreadable. Your users will have to keep zooming in and out to find what they want to read, and your links will be harder to click.

We grabbed the following image from Slate.com, one of our favorite online publications. Although their content is amazing, they haven't optimized their site for mobile users yet. You'll find this to be the case with a lot of websites. To build for mobile is to be on the cutting edge.

Everything on this site is squished into the frame. When people have a hard time using a website (or any piece of software), they tune out fast. It doesn't take too long before they're gone forever, off to check their Facebook status, off to watch hilarious cat videos on YouTube. With an easy-to-use website that's designed for mobile devices, none of this would be happening.

Here's why. Mobile websites make it easy on your users. You can shrink down your website so people can get to the content they want without having to dig around for it. It basically comes down to this. If you don't build a mobile version of your website, you won't capture the mobile market. It's as simple as that. It's not an option. It's a necessity.

Where do I start?

We'll want to start by separating your mobile website from your standard website. This can be done in a variety of ways. You can create your own mobile sub-domain for your website, or you can make some modifications to your website's .CSS files for the same effect.

Your mobile website should never interfere with your standard website. You don't want your standard website users to stumble on your mobile website if they aren't accessing it from a mobile device. Why? Because if they do, it will appear as though they've stepped

13 years into the past. Mobile websites have to be small to fit small screens, and that means most of them will look a lot like the internet of 1997.

We don't want to this to happen, so we'll create a mobile sub-domain on our server.

Create Your Own Mobile Sub-Domain

Have you ever been to a website that starts with a "m." ? Facebook is a good example. When you login to Facebook from a mobile device, you get redirected to "m.facebook.com," Facebook's mobile sub-domain. A mobile sub-domain allows you to separate the two designs you're using for your standard website and your mobile website, making it easier for you to create content specifically for your mobile users.

It's really easy to setup a mobile sub-domain. We'll go through the process in Cpanel, using a hosting account from Bluehost.com, but any hosting account that gives you access to the same Cpanel software will work. Once you're in Cpanel, go down to the domains and click on the subdomains icon. It looks like this.

Inside of the subdomain manager, you can easily create a ".m" sub-domain for your current website. You can also create your own folder where you'll store all of the files associated with that domain. If you plan on using a content management system like Wordpress or Drupal, this is where you will do your main install.

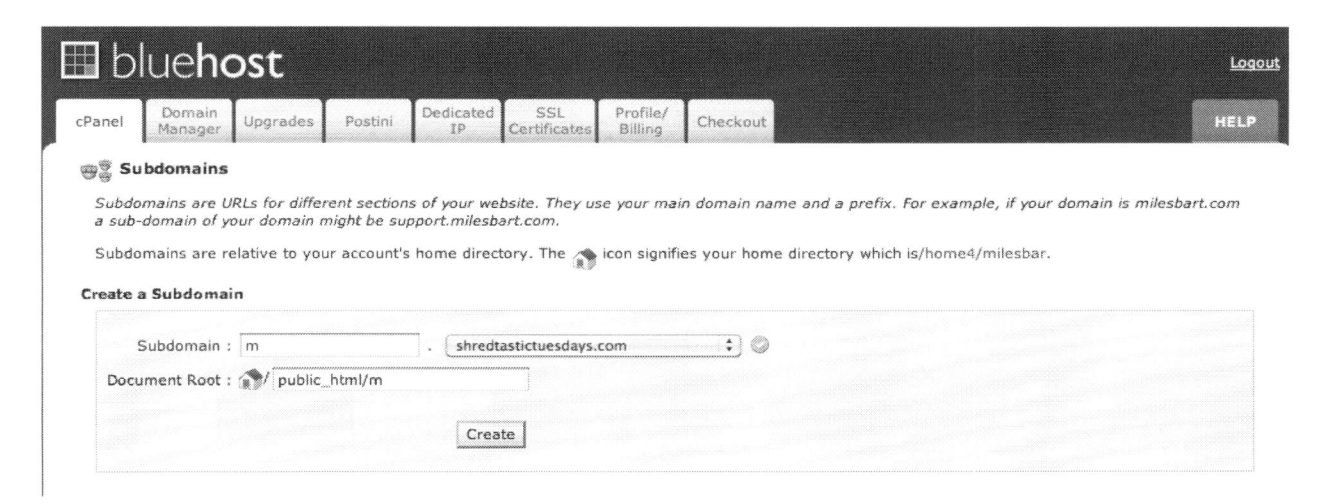

Redirect Users to Your Mobile Sub-Domain

There are a lot of ways to accomplish this task with different programming languages. Because we're sticking with Javascript and CSS in this guide, we're going to pick the

Javascript method. You can place the following piece of Javascript code in between the <head> and </head> tags in your index.html or index.php file. You can also place it in between the <body> and </body> tags on your individual pages to redirect your users to the mobile version of that page.

Here's a screenshot of how it should look:

```
1  <head>
2
3  <script type="text/javascript">
4  if(screen.width < 600){
5  window.location = "http://m.shredtastictuesdays.com";
6  }
7  </script>
8
9  </head>
```

This code does a few things. It detects your user's screen size, and if it finds the screen size to be below a certain number (namely 600 pixels), it redirects your user to your mobile sub-domain.

Just so you know, this code doesn't work for all smartphones. It only works for the ones that have Javascript enabled. Thankfully, that includes most of today's models, and it's certainly going to be true for the mobile devices of tomorrow. Unfortunately, you have to pick who you want to please, and if you're trying to cater to everyone, you'll be spending a lot more time developing your app. We think this is the easiest way to go about redirecting your users to your mobile sub-domain.

Very important! Make sure you do this!

If you are going to use Javascript redirects, you need to use them on *every single page of your website!* No user wants to be redirected to the main page when she wanted to access your blog post on funny cats. That makes for a horrible user experience. Here's how you can redirect your users to the mobile version of the same blog post or subpage.

Just place the redirect code somewhere in between your <body> and </body> tags. I did the following redirect on one of my Wordpress websites, using the post editor.

Edit Post

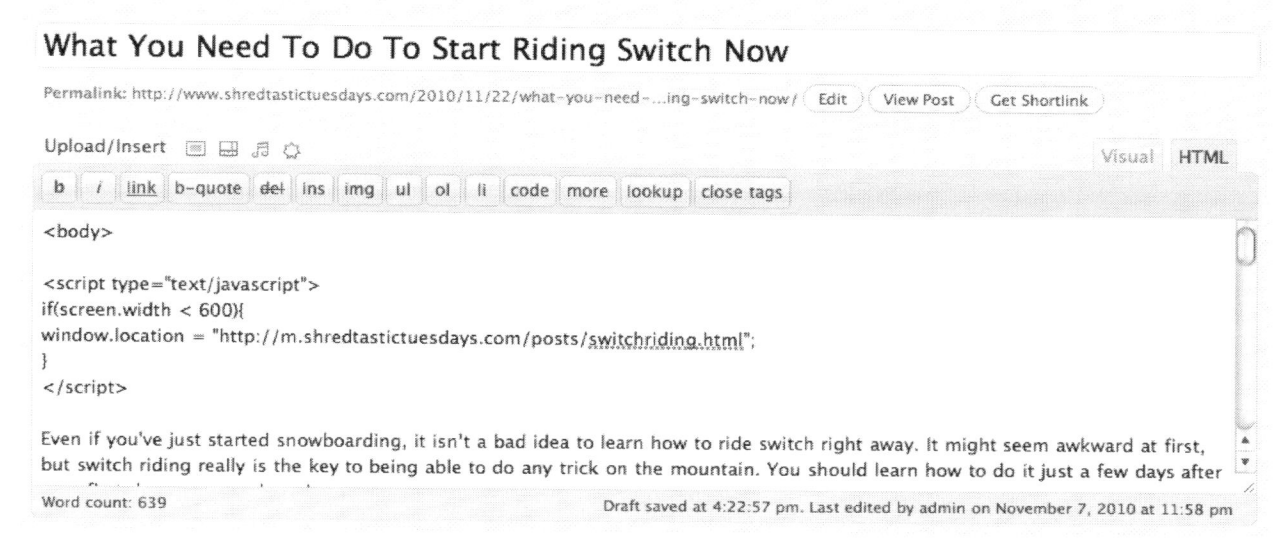

Now anybody who stumbles upon this post from a mobile device will be immediately taken to the mobile version of the post. Pretty cool stuff, and very easy to implement. You just have to remember to do it.

The Quickest and Easiest Way to Setup a Mobile Website

Once you've gotten your users to your mobile sub-domain, you'll need to have an actual website on that sub-domain. This can be done in very short order, using a mobile-specific Wordpress theme. You won't need to know anything about CSS or Javascript, and everything will be ready within minutes. Granted, your design won't be 100% unique, but you can always tweak it later or hire someone else to tweak it for you.

You can find dozens of mobile Wordpress themes by simply typing a query into Google. Once you've found one you like, download it to your computer and then upload it through your Wordpress theme manager on your mobile domain. That's the easy part.

Mobile Search Engine Optimization (SEO)

So what's the hard part? Well, now you've got two different websites with the exact same content. Every time you post something to your standard website, you'll need to post it again on your mobile website. It's a total pain, but there is no better way to solve this problem.

Unfortunately, it gets worse. You probably want your website to rank well on Google. Most of us do. But here's the problem. When you have the exact same content on multiple domains, Google assumes that your other domain is attempting to copy the content on your first domain, and it places your new content in the "duplicate" category.

When your pages are considered duplicates of another website's content, they are practically guaranteed to get no traffic from a search. That's a very bad thing, and there really is only one solution. *Create completely different written content for your mobile website.*

From a search engine optimization perspective, this is a good idea. Research shows that people who search from mobile devices tend to enter shorter search queries, and they don't want to read long drawn out articles. Imagine that most of your mobile audience is just 5 minutes away from that next meeting, and you'll get a good grasp of the content strategy you need to employ.

So, whenever you create a new page, a blog post, a video, or any piece of content for your standard website, think about how you can change it for your mobile website. You'll probably want to make it shorter while using less keywords in the title. Yes, you'll still need to rewrite everything from scratch, but it will be considerably easier. Brevity is a virtue in the mobile world. Keep everything under 500 words.

Don't want to use sub-domains and Javascript redirects? Try some simple changes in CSS.
CSS is a powerful tool for changing your entire website design in a matter of seconds. It also allows you create a completely different, mobile-specific design for your website without having to use sub-domains or any redirects. It isn't the best strategy for SEO purposes, but it can save you a lot of time. And, as an added bonus, you won't have to rewrite all of your website content for a mobile audience (unless you really want to).

We're going to introduce you to CSS, stylesheets, and Javascript. Some of you are probably experienced enough to know what all of these are. You've more than likely worked with them to create websites in the past. If this sounds like you, go ahead and skip to the next section where we show you how to use them to create your mobile design.

So, what is CSS? We'll be the first to say it's one of the more awesome things that's happened to the web. You see, in the early days of the internet, website designs and website content were an intertwined mess. If you wanted to change the font on all of your website's subpages, you actually had to go in there and manually enter HTML code into every ".html" page.

CSS is a game changer because it separates your website's design from its content. You can use it to specify a single style for your entire website, and that style will carry over to all of your pages. You can also use it to define a style that only applies to mobile devices. That's exactly what we're going to do.

You've probably seen files like "stylesheet.css" and wondered what they do. Simply put, they define your fonts, heading and subheading size, padding on the edges, and anything else that contributes to the look and feel of your website. A ".css" file is a basic recipe for your website's overall design.

O.K. Enough theory. What does this actually look like? Let's start with a few simple changes to a basic Wordpress site. Wordpress allows you to modify your site's CSS from its control panel, something that will definitely come in handy because it means you won't have to keep downloading and uploading your CSS files from your server.

Here's a snapshot from the website we'll be using for the example. It's a ski and snowboarding site that I setup in my spare time called ShredTastic Tuesdays. This is what the site looks like without any changes to its ".css" files.

15 dec 2010 | What To Do At Breckenridge Colorado

Breckenridge is one of the more legit places to ride in Colorado. In fact, we like it so much that we live and ride there all of the time. All in all, the park is the best thing about Breck. Hands down, they tend to have the best built jumps in all of Colorado, and we aren't the only people you will hear saying that. Last year, Freeway was blowing up and Park Lane was one of the best places you could go to progress.

Brian Locke throwing down in Freeway Terrain Park.

Photo taken by Ted Bendixson.

Breck Is, By Far, The Best Place To Progress Your Tricks.

Seriously, one of the best ways to take your tricks to bigger jumps is to head over to Breck. That's because you can hang out all day in Park Lane, dial something on those jumps, and then head straight to the Freeway line to really make a statement. I can't tell you how many tricks I've taken to the big line by doing this. That's because every jump in the series is designed to be just a little bigger than the one before it. If you can do something smoothly off the last jump in Park Lane, you can totally do it on the first jump in Freeway. You just have to believe in yourself.

You can get to your ".css" file editor in Wordpress by clicking on the "editor" link under "appearance." Here's what it looks like.

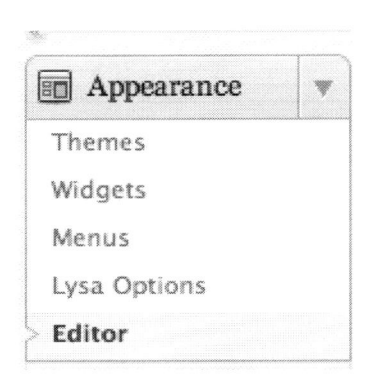

Once you're in the editor window, you'll see a bunch of different ".css" files under the "Styles" heading. We're going to edit "style-red.css" because it's the one that applies to the theme we're using for this site. Here's another snapshot.

When you click on the .css file, you get an editor window like this one. This is the actual ".css" code that defines your website's style.

style-Red.css

```
/* basics */

body {
        background:#780b12 url(images/red-bg.jpg);
        }

#container {
        color:#333;
        }

#box {
        background-color:#e9e9e9;
        }

a{
        color:#94090D;
        text-decoration:none;
        }

a:hover{
        text-decoration: underline;
        }
```

Cool, eh? If you change any attribute in this file, it applies to your entire website. Don't believe me? Watch what happens when I change the "background-color" attribute in the #box section. Here's what the code change looks like:

```
#box {
        background-color:#FFFFFF;
        }
```

What do you think this will do? It's definitely going to change something on the site, so let's have a look at the same screenshot after the CSS change.

15 dec 2010 | What To Do At Breckenridge Colorado

Breckenridge is one of the more legit places to ride in Colorado. In fact, we like it so much that we live and ride there all of the time. All in all, the park is the best thing about Breck. Hands down, they tend to have the best built jumps in all of Colorado, and we aren't the only people you will hear saying that. Last year, Freeway was blowing up and Park Lane was one of the best places you could go to progress.

Brian Locke throwing down in Freeway Terrain Park.

Photo taken by Ted Bendixson.

Did you catch that? One of the backgrounds changed over to white. Now, because this is a change in your ".css" stylesheet, it applies to every page on your website. Here's another screenshot from a different post.

What You Need To Do To Start Riding Switch Now

22 NOV 2010 BY ADMIN, NO COMMENTS »

Even if you've just started snowboarding, it isn't a bad idea to learn how to ride switch right away. It might seem awkward at first, but switch riding really is the key to being able to do any trick on the mountain. You should learn how to do it just a few days after your first day on a snowboard.

Wanna do this trick? You're gonna have to learn switch. Rider, Ted Bendixson.

Many thanks to Dan Weems for snapping this photo.

Wow! We modified *one teeny tiny piece of code*, and it changed the background on the entire site. That's pretty powerful stuff. Can we change more? You bet we can! Let's give it a try.

What if we don't like the headline color? What if we want it to be red to match the rest of the site? Well, you can easily change that in CSS as long as you know where to find it. Let's look around.

To start, I did a quick "view source" on the page to find out which tags Wordpress uses for my headings. As it turns out, Wordpress uses the h2 tag. To change that tag's color, I need only find its style definition in my main "style.css" file. Here's another snapshot from that file.

Stylesheet (style.css)

```
*{margin:0; padding:0;}


body {
        font-family: verdana, arial, tahoma, sans-serif;
        font-size: 9pt;
        margin:0;
        line-height:16pt;
        }

h1 {font-size: 24pt; font-family:Arial, Helvetica, sans-serif;}
h2 {font-size: 20pt;font-family:Arial, Helvetica, sans-serif; font-weight:bold; padding:10px 0;}
h3 {font-size: 16pt;font-family:Arial, Helvetica, sans-serif; font-weight:normal;}
h4, h5, h6 {font-size: 14pt;font-family:Arial, Helvetica, sans-serif;}
```

You can see the default fonts, font sizes, margins, padding, and bunch of other style elements that define the website's design. Do you see where "h2" is defined? Right now, no color has been set. So let's add the color red right in!

Here's what the modified h2 definition looks like when we want to change the headline to red:

```
h2 {font-size: 20pt;font-family:Arial, Helvetica, sans-serif; font-weight:bold; padding:10px 0; color:#ff0000;}
```

Do you see the end where color is specified as "#ff0000;" That's nerd-speak, or hex, for the color "red." After we've made the change, our website looks like this.

What You Need To Do To Start Riding Switch Now

22 NOV 2010 BY ADMIN, NO COMMENTS »

Even if you've just started snowboarding, it isn't a bad idea to learn how to ride switch right away. It might seem awkward at first, but switch riding really is the key to being able to do any trick on the mountain. You should learn how to do it just a few days after your first day on a snowboard.

Wanna do this trick? You're gonna have to learn switch. Rider, Ted Bendixson.

Many thanks to Dan Weems for snapping this photo.

What Is Switch Riding?

FOLLOW ME!

RSS feed   Twitter   facebook   flickr

PAGES

About

ARCHIVES	META
December 2010	Site Admin
November 2010	Log out
September	Valid XHTML
2010	XFN
	WordPress

O.K. So the red doesn't match perfectly. But given enough time, you should be able to come up with a more toned-down red to match the rest of your site.

If there's one point I want to make about CSS, it's this. *Don't reinvent the wheel when you're learning CSS.* It's not an entirely new programming language or anything like that. It's more like a website design control panel. You're much better off learning it by modifying an existing website than by creating ".css" files from scratch. I'm speaking from experience here.

Do what I've done. Get a hosting account, install Wordpress on your domain, pick a theme, and then modify that theme's CSS directly from your Wordpress control panel. Believe me, it's much easier that way. You'll learn it faster, and you'll have a working website to show for it.

Chapter 3: Creating a Mobile Website with CSS

Earlier, we were able to change our website's design by tweaking a few settings in the "style.css" file. Now we're going to create an entirely new design for mobile devices, and we're going to do it all from our main "style.css" file. Here's how you can get started.

1.) Define the section where your mobile design will reside. Some people like to put this at the end of their stylesheet, but any place you'll remember works just fine. Make sure you write a comment so you don't forget where your mobile design starts. Here's a screenshot with an example, using the same Wordpress site from before.

```
                    ...gin: 2px;
    }


    /* Mobile Design Definition */

    @media handheld{

            /* Our mobile design resides in these brackets */


    }
    /* other */
```

2.) Use "@media handheld" to create a design that only applies to mobile phones. Everything in the brackets after "@media handheld" will only show up on a mobile device. It's like having a separate stylesheet for mobile phones.

3.) Begin to define your website design for mobile phones. Once you've started, you need to redefine all of the fonts, headings, sub-headings, spacing, and other style elements for your mobile design. You can copy most of this information from your original "style.css" file. When you do this, your CSS should look something like the next snapshot.

```
@media handheld{

        /* Our mobile design resides in these brackets */
        body {
        font-family: verdana, arial, tahoma, sans-serif;
        font-size: 9pt;
        margin:0;
        line-height:16pt;
        }

        h1 {font-size: 24pt; font-family:Arial, Helvetica, sans-serif;}
        h2 {font-size: 20pt;font-family:Arial, Helvetica, sans-serif; font-weight:bold; padding:10px 0; color:#ff0000;}
        h3 {font-size: 16pt;font-family:Arial, Helvetica, sans-serif; font-weight:normal;}
        h4, h5, h6 {font-size: 14pt;font-family:Arial, Helvetica, sans-serif;}

        img{ border: none; padding:0; overflow:hidden;}
        img a{border:none;}

}
```

4.) Tweak and test. Get out your iPhone, and go to your website. Keep making changes to your CSS in the @media handheld brackets, and take note of its effect on your design. Keep tweaking until you like what you see.

What CSS Changes Are the Most Important for Mobile Design?

Step 4 is a good idea in principle, but it doesn't work unless you have the proper direction. How do you know what to change? How do you know what to keep the same? Here are a few things we've learned throughout the years.

1.) Keep it simple. Mobile visitors only want to see your content. Your mobile design should strip away all of the bells and whistles that go with a standard website. You don't need a sidebar. You don't need a big header or footer. You just need one column with your content.

You can tell CSS to avoid displaying the header, footer, or sidebar with the following piece of code:

```
/* Mobile Design Definition */

@media handheld{

        /* Our mobile design resides in these brackets */

        /* Get rid of the sidebar and footer */
        #sidebar, #footer {
                display: none;
        }

}
```

2.) Set a maximum width for all of your images. Remember, we want to get rid of all the pinching and horizontal scrolling. When your website has large images, your users inevitably have to scroll left or right to see them. Let's get rid of that by adding the following bit of CSS code.

```css
/* Constrain images to 250 pixels wide */
#content img {
        max-width: 250px;
}

}
```

Now all images larger than 250 pixels will be automatically scaled to 250 pixels when someone visits your site on a mobile phone. Easy.

3.) Get rid of excess padding. Mobile websites can't be as spread out as standard websites, so we'll need to get rid of some padding and margins. You can modify the "html" and "body" elements to accomplish this.

```css
/* Get rid of padding and pick a normal font */
html, body {
        padding: 3px;
        margin: 0;
}
```

We haven't gotten rid of all the padding. We kept a little bit so your website doesn't look completely cramped.

4.) Get rid of distracting background images. They're nice to have on a fully-featured laptop or desktop machine, but they only slow down your mobile website's load time. Welcome to the world of 1997. Connection speed is still an issue with smartphones. We've got at least 3 or 4 more years before every mobile phone catches up to high speed cable internet.

Here's the code. You're simply going to change the background to a solid color of your choosing. This code comes right after your padding and margin changes in the html and body elements. If you place it outside of those elements, your background color won't change.

```
/*Get rid of padding */
html, body {
padding: 3px;
margin: 0;

/* Change background to a solid color */
background:#fff;


}
```

5.) Scale down your headings so they don't overpower your users' mobile devices.
Big headlines can really make your content stand out when you're displaying it on a
laptop or desktop machine. Unfortunately, they're a little *too* big when you're viewing
your website from a mobile device. Here's some code you can use to scale all of your
headings down.

```
/* Scale down headings so they don't explode on the page */
h1, h2, h3, h4, h5, h6 {
font-weight:normal;
}
```

6.) Don't forget to make your links stand out. You've just changed your entire design,
so your background is unlikely to contrast with your old text colors. Better make sure you
pick the right link colors so they stand out and get clicked. Here are some bits of CSS
code you can use to redefine your link colors.

```
/* Don't forget about links. They need to stand out too. */
a:link, a:visited {
text-decoration: underline;
color:blue;
}

a:hover, a:active {
text-decoration: underline;
color: #660066;
}
```

You're probably wondering why CSS doesn't refer to the link element as "link." This is
actually an old convention from HTML. You've no doubt used "a href" before. That's the
tag we're modifying when we set the above attributes in CSS.

There are four of them, each representing a different link state. "Link" refers to a link that
has not been visited. "Visited" is pretty obvious in light of this. It refers to links your
users have visited. "Hover" designates what will happen to the link's color when the
mouse pointer hovers over it, and "active" is like a switch your can activate in other parts
of your html code.

Because most mobile users are visiting your website from touchscreen devices, it's important to remove all of the bells and whistles from your links. A "hover" state never really occurs on a touch-based device, so it's unwise to give a style that's too different from the "link" and "visited" states. In the above code, we only changed the color. You'll want to avoid any changes in text decoration or text size.

A Few More Options to Consider

These 6 changes represent the most basic steps you need to take to optimize your design for mobile phones. Where you go from there is completely up to you, and you'll only know by trying out different design options. You can, for example, define a completely unique design for the iPhone by entering the following code.

```
/* This design only applies to the Apple iPhone 4 */

@media only screen and (max-device-width: 640px){

        /* iPhone 4 design code goes here */

        html, body{
        background: #f00;
        padding: 5px;
        margin: 0;
        }

}
```

The above code only applies to devices with screens that have a maximum width of 640 pixels. Right now, that's the maximum width of the iPhone 4. Okay, so we lied. It's also the maximum width of a few other mobile phones and some very old desktop computers. But we're pretty sure those machines don't even understand CSS, so they're out of the picture.

Creating a Mobile Stylesheet and Link to It On All of Your Pages

If you want your mobile stylesheet to be completely separate from your main stylesheet, you can create a mobile stylesheet and link to it on all of your pages. When you are running a Wordpress-based website, this is very easy. You simply add something like the following piece of code to your "header.php" file.

```
<link rel="stylesheet" href="http://www.shredtastictuesdays.com/mobile.css" type="text/css" media="handheld">
```

We're using the same example website from before. Under the "href" attribute, you'll see the location of our mobile stylesheet. That's the ".css" file that will contain the same mobile design information we used above. The "media" attribute is also very important. We set it to "handheld" just like we did in the main stylesheet earlier.

When you link to a stylesheet and you specify the media type in the link tag, you don't have to use the "@media { }" separator in your mobile stylesheet. You simply write it like you would write any other stylesheet you're using.

Where to Place to Your Mobile Stylesheet Link

When you place the link code in your "header.php" file, it effectively links to your mobile stylesheet on every page of your website. You'll want to add your mobile stylesheet link right after your standard "screen" links. These can be found in your Wordpress "header.php" file, using the same appearance->editor we've been using to modify your site's CSS. Here's another screenshot from the editor, just in case you're having some difficulty finding it.

It's right there, under templates. If you squint, you can see the link code above, but here it is again below.

```
<link rel="stylesheet" href="<?php bloginfo('stylesheet_url'); ?>" type="text/css" media="Screen" />
<link rel="stylesheet" href="<?php bloginfo('stylesheet_directory'); ?>/style-<?php echo $pov_color; ?>.css" type="text/css" media="Screen" />

<link rel="stylesheet" href="http://www.shredtastictuesdays.com/mobile.css" type="text/css" media="handheld">
```

Notice how the "Screen" media type is capitalized. That's no accident. You need to do that in order to make your website viewable on Windows Mobile handsets. Once you've done this, you simply need to create your mobile.css file using the same code we used earlier. Once you've uploaded that file to your server, you're done.

Javascript Redirects and Mobile Wordpress Themes vs. CSS. Which is Best for You?

It depends on your website content strategy. If you are simply creating a business website with some very basic information about your company, what you do, and how people can get in contact with you, go with CSS. You'll find it much less complicated, and it won't involve nearly as much work. That's because your mobile audience won't differ all that much from your desktop/laptop audience. You're not giving them a novel. You're giving them a synopsis.

But, if you've got a big blog, and you're always writing useful content to help your audience, you'll probably want to go with a mobile sub-domain and an entirely different mobile website. That's because a large part of your business involves getting traffic from search, and that means you'll need to tailor your blog posts to your mobile audience. A few simple changes in CSS won't do that for you, but an entirely different mobile website with a mobile-specific Wordpress theme will.

So, ask yourself this question. Who is your audience? Do you need to create a separate mobile experience to please them, or can you do the trick with a few simple changes to your already existing website? How much work can you afford to put into this?

Chapter 4: Using Javascript to Make Your mobile Website More iPhone-like.

CSS isn't the only tool you can use to build a fully functional mobile website. With Javascript and HTML, you can create a more interactive and animated user experience. When you do it right, your users won't even know the difference between your mobile app and an iPhone app they've downloaded themselves.

We're going to explore just one of many options you have at your disposal when working with Javascript. This particular Javascript library mimics the iPhone's menus, back buttons, animations, and all other user interface elements that separate the iPhone from every other handset. It's called iUI, and it's an excellent springboard for the rest of your mobile development. Let's have a look under the hood.

You can download iUI by going to the following address:
http://iui.googlecode.com/files/iui-0.31.tar.gz

Once you unpack iUI, you'll get a folder containing some very iPhone-esque graphics as well as the javascript and CSS files themselves. For the purpose of this tutorial, we will only be working with iUI.js and iUI.css, the other files iUIx.css and iUIx.js are the compressed versions.

What you should see in you iUI folder.

Go ahead and upload your iUI folder to your server. We're going to get started by building a simple mobile website.

How to Link to Your iUI Library

Create your index.html file in the code editor of your choice. The cool thing about iUI is that it doesn't require that much knowledge of Javascript. You only need to know enough to link to the right files and to use them with HTML to build your site. Once you've done that, you simply let the Javascript library do the work for you.

Here's what the first part of your index.html file should look like. We'll be linking to the Javascript and CSS libraries within the header element.

```
1   <html xmlns ="http://www.w3.org/1999/xhtml">
2   <head>
3   <title>Sample iUI Mobile App</title>
4
5   <meta name="viewport" content="width=device-width; initial-scale=1.0; maximum-scale=1.0; user-scalable=0;"/>
6   <style type="text/css" media="screen">@import "http://www.shredtastictuesdays.com/iui/iui.css";</style>
7   <script type="application/x-javascript" src="http://www.shredtastictuesdays.com/iui/iui.js"></script>
8   </head>
9
10  <body>
11
```

You can see that I only need to upload the "iUI.css" and "iUI.js" files in order to use this particular Javascript library. I'm also setting my meta viewport element to reflect the small mobile phone screen size. This should be pretty easy as long as you know where you've placed your files.

How to Make a Mobile Web App with iUI

Here's where it gets fun. iUI creates a whole set of div classes that you can use to build iPhone-like apps. If you aren't familiar with div classes, don't worry. I'll explain them in this section. They're basically a web design element that you create using HTML, kind of like a layout divider (hence the name).

To use iUI, we simply refer to the div classes defined in the iUI Javascript library that we just uploaded. You'll be pleasantly surprised with how easy the entire process is. I know I was. So let's get started.

The next piece of code creates the classic iPhone-style toolbar at the top. The div class we're using from our Javascript library is called the toolbar class. Inside of it, we can use some more familiar html elements like "h1". Have a look.

```
<body>

<div class="toolbar">
    <h1 id="pageTitle">Skiing</h1>
    <a id="backButton" class="button" href="#"></a>
    <a class="button" href="#searchForm">Search</a>
</div>
```

I've also got two "a" elements. The first one basically creates the back button's functionality, using the # href to refer to the previous page. The next one creates an iPhone-like button that links to your web app's search function (also programmed into the Javascript library you've installed).

If we were to stop right here, this is what your iPhone web-app would look like.

Not too shabby for not doing much hard coding at all. You'll notice the very iPhone-esque search button as well as the title. Your users will believe they're using a native app, even though they're just visiting your website.

Now we're going to build the menu options below. iUI basically hijacks some of your favorite HTML elements like "ul" and "li." If you've used these before, you know that they usually create a bulleted list. Now you're going to use them as your menu options.

```
<ul id="home" title="Skiing" selected="true">
    <li><a href="#mountains">Mountains</a></li>
    <li><a href="#proskiers">Pro Skiers</a></li>
    <li><a href="#bigtricks">Big Tricks</a></li>
    <li><a href="#shredstats">Your Shred Stats</a></li>
    <li><a href="http://m.shredastictuesdays.com/about.html" target="_self">About</a></li>
    <li>Some Text</li>
</ul>
```

The key to doing to this right is setting the "ul" id to "home." This is what your Javascript library recognizes and uses to create a submenu with the look and feel of an iPhone app. You'll notice that I've given each menu option its own href such as "#mountains," "#proskiers," etc. We'll refer to these later as we expand our app.

The second to last <li> element is a link to the site's about page, and the last element is just some text. This is merely meant to show you a few things you can do with this app.

Here's what your app should look like now:

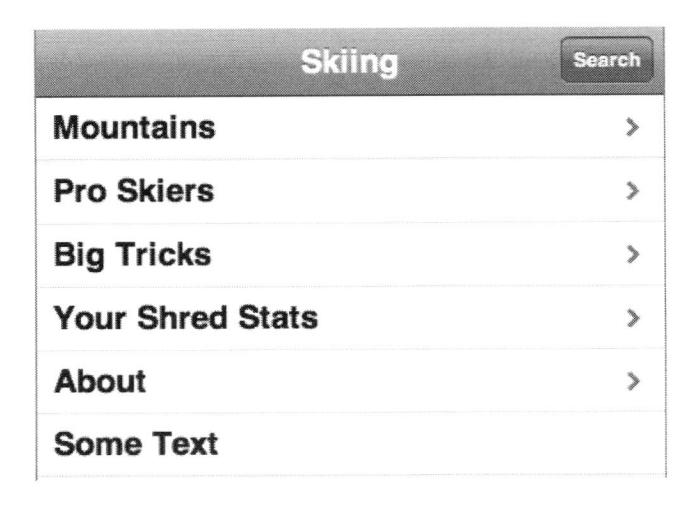

If you're a pretty seasoned coder, you can probably get to this point within one hour. That's actually pretty amazing. It looks just like a real iPhone app.

So where do we go from here? Well, we can start out by building in some submenus so our app actually does something when you click on the links. To create a submenu, you're basically going to do the same thing you did to create the menu above. The only difference is that, this time, we're going to reference the "#mountains" list id. Here's what that looks like in code.

```
26
27  <ul id="mountains" title="Mountains">
28      <li class="group">A</li>
29      <li><a href="#abasin">A-Basin</a></li>
30      <li class="group">B</li>
31      <li><a href="#bigsky">Big Sky</a></li>
32      <li><a href="#breckenridge">Breckenridge</a></li>
33      <li class="group">C</li>
34      <li><a href="#copper">Copper</a></li>
35      <li class="group">K</li>
36      <li><a href="#keystone">Keystone</a></li>
37      <li class ="group">M</li>
38      <li><a href="#meadows">Mt. Hood Meadows</a></li>
39      <li class ="group">T</li>
40      <li><a href="#timberline">Timberline</a></li>
41  </ul>
42  </body>
43  </html>
```

You'll start noticing that iUI has a bunch Apple-style designs elements that you can easily use. One of them is the "group" element, something you'll remember from iTunes. The code we just used will generate an alphabetical list of different mountains where you can go skiing. Here's a picture of the design you'll get from using this code.

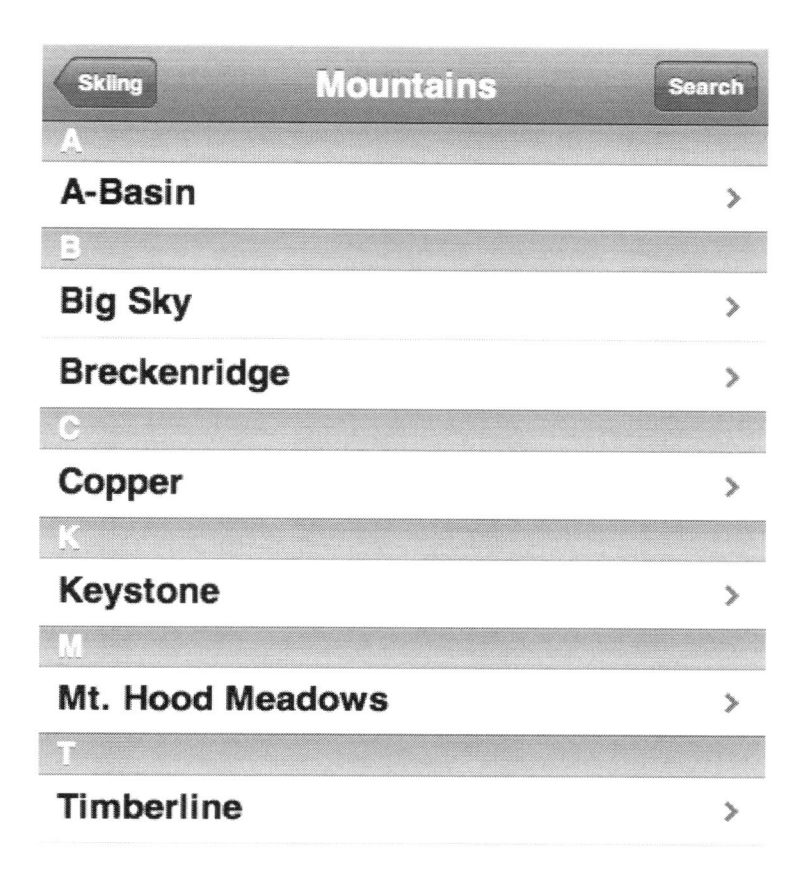

Not bad, I must say. Did you notice the back button up top? When your users tap on it, they'll be taken back to the original skiing page.

By now, you're probably wondering if we're ever going to add any real content to the app. Of course we are! What would be the point if we weren't? Let's say someone clicks on "Breckenridge" (You must ski there BTW), and we want to tell them a little bit about it. How might we go about doing that? Let's have a look.

```
<div id="breckenridge" class="panel" title="Breckenridge">
    <h2>Breckenridge is one of the best places to ride park. There's jumps...</h2>
</div>
```

You'll remember earlier that we assigned "#breckenridge" to the Breckenridge menu item. Now we're using that same name in our new div. That way, when you click on "Breckenridge" in the menu, you get this screen.

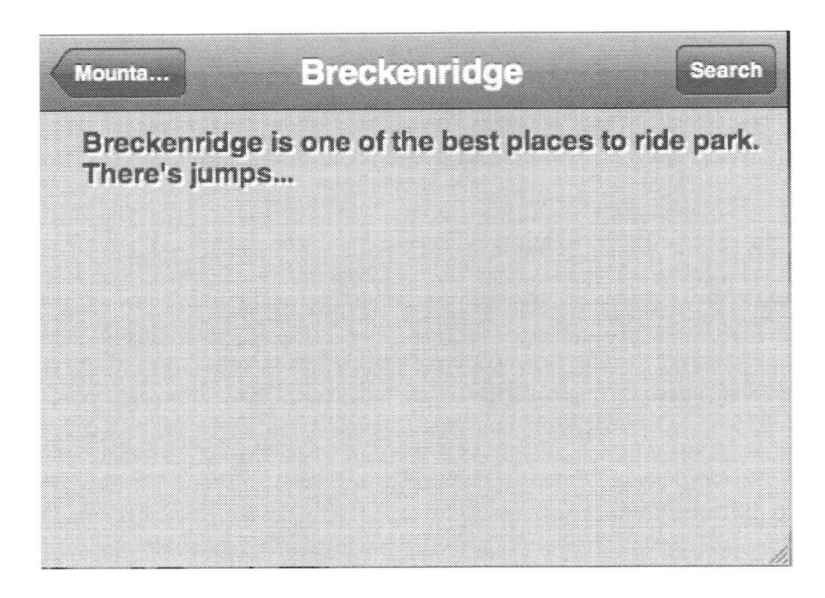

Very cool. I didn't go through the liberty of filling up the whole frame, but you could use this form for all kinds of things. If you add an <img> tag in there, you can use pictures just like any other mobile website.

```
<div id="breckenridge" class="panel" title="Breckenridge">
    <h2>Breckenridge is one of the best places to ride park. There's jumps...</h2>
    <img src="http://www.shredtastictuesdays.com/iui/tedcork5freeway.jpg" width="300" height="240"></img>
</div>
```

It's not perfect, but with some centering, you can make it look good.

Oh, and just one word of advice. If you're going to use picture content with iUI, you'll need to resize it to fit a mobile phone screen. Otherwise, iUI will literally freak out on you. It will constantly attempt to resize itself to fit the image, making your app practically unusable. Be careful.

Use iUI to Create a Login Screen for Your App

And you thought we were finished. Not a chance. You can use iUI for all sorts of things. Now we're going to use it to create a login screen. This involves putting a bunch of different pre-programmed iPhone UI elements together. We're going to use the "row" div and "white button" class. With these, your login screen will look just like the real thing.

Let's imagine that you want to see your Shred Stats. Everyone has different Shred Stats, so you'll need to login to see yours. Here's the code you would use to build a login screen with iUI.

```
<form id="shredstats" title="Shred Stats" class="panel" method="post" action="login.php" target="_self">
    <fieldset>
        <div class="row">
            <label>Username</label>
            <input type="text" name="username" value=""/>
        </div>
        <div class="row">
            <label>Password</label>
            <input type="password" name="password" value=""/>
        </div>
    </fieldset>
    <a class="whiteButton" type="submit" href="#">Login</a>
</form>
```

First things first, look at the form id. It references the "shredstats" menu item that we listed at the very beginning. That way, when you click on it, you'll be taken to the form. Next, we've got the "panel" class, the same thing we just used to display the short description and website content.

The rest of the code in the <form> tag refers to the way the login screen will actually function. "login.php" is the backbone of the system. It interacts with a database to check your username and password. Unfortunately, we don't have the time to cover the fundamentals of php, databases, and login screens in this section. We simply want to show you what's possible with iUI and Javascript.

When you're all done entering this code, you should get the following screen after clicking on "Your Shred Stats."

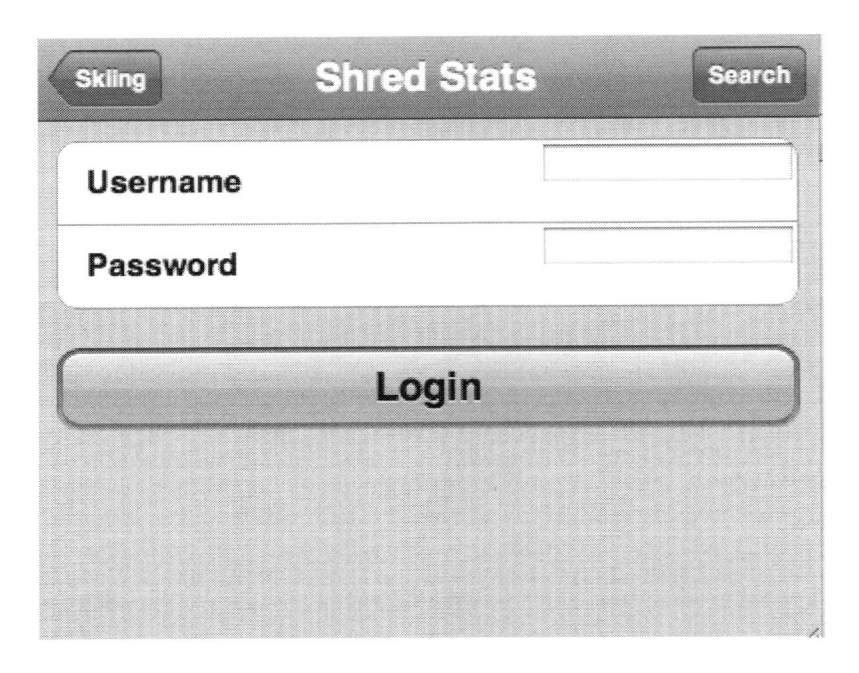

Once again, this is pretty amazing considering that you don't need to know a thing about Javascript. You only have to know enough to import the library and use these basic HTML elements. That's powerful stuff.

How to use iPhone-esque sliders to create a control panel that looks just like the real thing.
The guys who created iUI also added in the classic slider switch that is oh so characteristic of the iPhone. You can use it in your web app designs too. For the purpose of this exercise, let's say you want to create a toggle for email notifications, and you want to place that toggle on your ShredStats page. How might you go about it?

Well, you would need to reference the div "toggle" class, using spans for the on and off switch. The code is just below the code for the password box. Here's what it looks like.

```
<div class="row">
    <label>Password</label>
    <input type="password" name="password" value=""/>
</div>
<div class="row">
    <label>Get Notifications?</label>
    <div class="toggle" onclick=""><span class="thumb"></span><span class="toggleOn">ON</span><span class="toggleOff">OFF</span>
    </div>
</div>
</fieldset>
```

You can see how the on and off spans are right next to each other. With this code, you get a login screen that looks like this.

On an iPhone, this toggle switch works just like the real thing. It moves with your finger as you slide it across the screen.

Chapter 5: What Else Can You Do With Javascript?

Javascript is a fully-fledged programming language for making web apps, and we've only shown you the surface layer. Now we're going to show you a few ways to put some real functionality into your web apps. How are we going to do this? We're going to build a temperature converter app using the same iUI interface. Let's get started.

Traveling the world is awesome, but if you're from the U.S. like I am, you start to get really confused when other people start talking about the weather. Wouldn't it be nice to have a web app that does the temperature conversion for you, mid-conversation? Well guess what. We're going to add this feature to our Skiing app.

To get started, we need to build the form we'll be using for our conversion. If you've read the previous section then this should be all but completely familiar. We're going to get rid of the last bit of useless sample text at the end of the first menu, and we're going to place a link to our temperature converter just before the "about us" page link.

Here's the code for that.

```
<ul id="home" title="Skiing" selected="true">
    <li><a href="#mountains">Mountains</a></li>
    <li><a href="#proskiers">Pro Skiers</a></li>
    <li><a href="#bigtricks">Big Tricks</a></li>
    <li><a href="#temperature">Temperature Converter</a></li>
    <li><a href="#shredstats">Your Shred Stats</a></li>
    <li><a href="http://m.shredastictuesdays.com/about.html" target="_self">About</a></li>
</ul>
```

So now we have a link to #temperature. From here, we only need to create a form that's similar to the login form we used earlier. We'll split it into three sections. One of them will contain the temperature in Fahrenheit, another will show the same temperature in Celsius, and the third element will be a "convert" button you press to do the temperature conversion.

Here's the code for that:

```
<form id="temperature" title="Converter" class="panel">
    <fieldset>
        <div class="row">
            <label>Fahrenheit</label>
            <input type="text" name="ftemp" value="32"/>
        </div>
        <div class="row">
            <label>Celsisus</label>
            <input type="text" name="ctemp" value="0"/>
        </div>
    </fieldset>
    <a class="whiteButton" type="submit" onclick="ctemp.value=tempconvert(ftemp.value)" href="#temperature.ctemp">Convert</a>
</form>
```

Don't worry if you don't understand some of this. We'll be explaining it as we look at using Javascript to put some functionality into the app.

As you can see, we've got a Fahrenheit box with a default value of 32, the freezing temperature of water. We've also got a Celsius box set to 0, the freezing point of water in Celsius. The white button activates the Javascript function we'll be using to calculate our temperature.

Your iUI design should look like this when you're done:

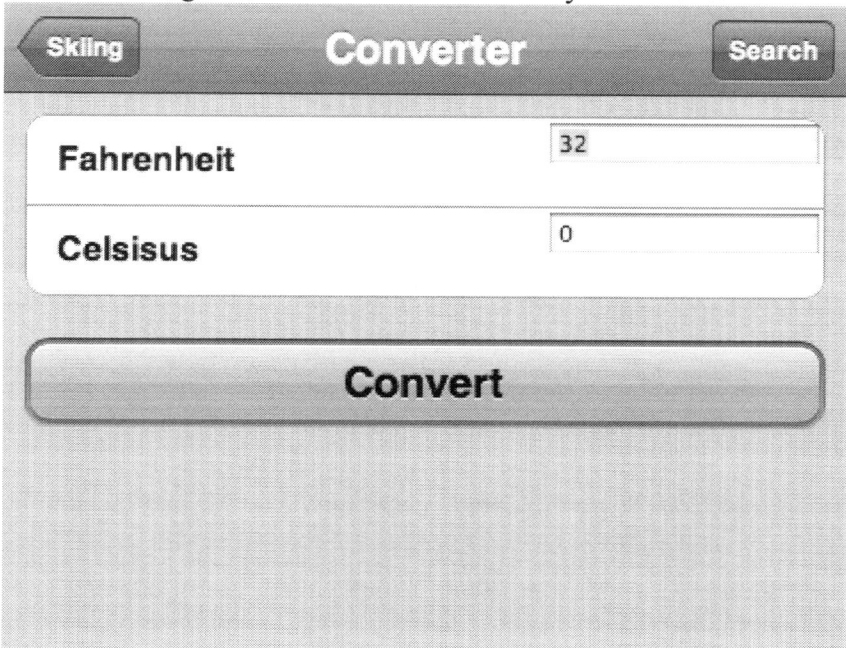

How to Write a Javascript Function

That was the easy part. Now we've got to figure out how to calculate the temperature conversion and get it to show up in your temperature converter program. This involves some general understanding of Javascript and a few computer programming fundamentals like passing values through functions. If you have some experience with this, it should be completely familiar. If you don't, I'll explain everything.

We're going to write a Javascript function to calculate the Fahrenheit to Celsius conversion. But before we can do that, we need to know where to put our function and how to use Javascript in an HTML document.

You can insert Javascript in between your document's <head> tags, <body> tags, or you can simply reference an external .js file. We've already referenced the Javascript file "iUI.js" for our iPhone-esque user interface, and we did that in the <head> tag. Now we're going to write some Javascript in between the <body> and </body> tags.

Here's what that looks like:

```
0 <body>
1 <script type="text/javascript">
2 /* Code goes here */
3 </script>
4
```

Our function will be called "tempconvert," and it will do the temperature conversion for us. Tempconvert takes in a number, calculates the Fahrenheit to Celsius conversion, and returns the converted number. We'll write it like this, in between our <script> tags.

```
1 <script type="text/javascript">
2 function tempconvert(ftemp){
3
4 }
5 </script>
6
```

ftemp is the number that the user will enter into the temperature conversion form. It is the temperature in Fahrenheit that we will need to convert to Celsius.

Some of you might remember that you can convert any number from Fahrenheit to Celsius by following these steps.

Take the Fahrenheit temperature and subtract 32
Divide this number by 1.8

Here are those two steps in our Javascript function:

```
2 <script type="text/javascript">
3 function tempconvert(ftemp){
4     ftemp = ftemp - 32;
5     ftemp = ftemp/1.8;
```

You can see that ftemp acts like a placeholder in each step. For step #1, we are setting it equal to itself minus 32. The change then carries over to the next step where ftemp is once again set to itself, but this time it is divided by 1.8.

After these two steps, ftemp should be the correct temperature in Celsius. But we want to display clean numbers, so we'll do a little rounding. In the next piece of code, we set ftemp to a rounded version of itself using the Math.round() Javascript function.

```
2 <script type="text/javascript">
3 function tempconvert(ftemp){
4     ftemp = ftemp - 32;
5     ftemp = ftemp/1.8;
6     ftemp = Math.round(ftemp);
```

"Math" is a collection of different Javascript functions that do all kinds of calculations. We called the round function, but there are plenty more where that came from. Math.random(), for example, returns a random number depending on a few settings you give it. You can look up all of these functions online.

As a final step, we're going to return our ftemp value so it can be displayed in our user interface. The complete function should look like this.

```
<script type="text/javascript">
function tempconvert(ftemp){
    ftemp = ftemp - 32;
    ftemp = ftemp/1.8;
    ftemp = Math.round(ftemp);
    return(ftemp);
}
</script>
```

Before we move on to the next part, I should warn you of a few rookie errors people tend to make when they're working with Javascript. One of them is forgetting to close off all functions with the ending "}" bracket. If you don't do this, your code might not work, and you'll staring at your screen for hours trying to figure it out.

Another one happens when you forget to place a semicolon at the end of each line inside of a function. It's a sneaky one because it's hard to spot, but it will still ruin your code.

How to Display the Results of Your Temperature Conversion Function

By now, you know how calculate your temperature conversion. But how do you get it to show up in your app? That is the point, after all. I'll admit that it took me a bit of time to figure it out, but it isn't as complicated as I thought it could have been.

We already know one thing. The "convert" button will set everything in motion. Let's have another look at the form containing our two temperatures.

```
6  <form id="temperature" title="Converter" class="panel">
7      <fieldset>
8          <div class="row">
9              <label>Fahrenheit</label>
0              <input type="text" name="ftemp" value="32"/>
1          </div>
2          <div class="row">
3              <label>Celsius</label>
4              <input type="text" name="ctemp" value="0"/>
5          </div>
6      </fieldset>
7      <a class="whiteButton" type="submit" onclick="ctemp.value=tempconvert(ftemp.value)" href="#temperature.ctemp">Convert</a>
8  </form>
```

Do you see the two input types? The one for Fahrenheit is named "ftemp" and the one for Celsius is named "ctemp" for some pretty straightforward reasons.

Interestingly, we can use these names to get at the numbers that the user types into the form. It's simply the name of the input form, plus a dot, and then "value." So...

ftemp.value = the number the user types in for the fahrenheit temperature.
ctemp.value = the temperature displayed in celsius.

All of the real work is done with the "onclick" statement embedded in the "convert" button code. Let's look at that piece a little more closely:

```
onclick="ctemp.value=tempconvert(ftemp.value)"
```

So, what exactly does this do? First, we know that it's setting the Celsius temperature (ctemp.value) to some number. How is it doing that? It's using the function we wrote, tempconvert, on the Fahrenheit temperature that the user entered into the form.

Tempconvert returns a temperature in Celsius, starting off with a temperature in Fahrenheit. So we know that ctemp.value will be set to a temperature in Celsius.

How Does All This Show Up On the Display?

Whenever you change the value of an input box, the user interface automatically displays the new value in that input box. That's how this piece of code works. We take the value from the form, feed it through a conversion function, and then we change the value being displayed in another input box. As soon as ctemp.value changes, it gets displayed in our app. It's as simple as that.

Well... not exactly. I had to do a few extra things to stop the app from crashing. One of them involves this piece of code:

```
href="#temperature.ctemp":
```

What on Earth does that do? You've probably noticed that we've used hrefs before. That's how we built our menu navigation system. But the other hrefs referred to categories and menus we've created for our app. This one doesn't seem to be doing that at all. In fact, it's only referring to a part of the form we're already using. What's going on here?

Ironically, we're using this bit of code to stop the "convert" button from taking us to another part of the app. It acts as a sort of "stop" function. That way, when you click on the convert button, you'll stay on the same page, and you'll see your converted temperature. Ingenious? Maybe. Total Kluge? Much more likely. There's probably a more simple way to do it, but this piece of code works for now.

So go ahead give this thing a try. It should work flawlessly. And, as an extra credit assignment, try to build a Celsius to Fahrenheit converter. It's certainly a little more complicated, but it's not entirely out of your reach.

Wrapping things up...

In this chapter, we've shown you why mobile websites are different from standard websites and how to create one of your own. We played around with CSS and Javascript, showing you a few ways you can create fully functional iPhone-esque web apps without having to reinvent the wheel.

But wait a minute. This guide isn't about web apps. It's about PhoneGap. In the next section, we'll show you how to use PhoneGap to turn your web-app into a native iPhone App. We'll introduce the iPhone SDK, and we'll help you solve some of the common problems developers encounter when they build their first iPhone app.

Here are some parting thoughts. When building your mobile app, don't do everything from scratch. Find a handy javascript library, a nice Wordpress theme you like, or some other template, and then work from there. You can always hire someone to take care of the pieces of code you don't understand. You'll waste a lot less time this way, and the bulk of your work will be done before you even start.

I'll say it again. Never, ever, reinvent the wheel. Someone has solved your problem before. Look for that solution before you attempt anything else. Happy coding!

Chapter 6: How To Turn Your Web App Into An iPhone

Now that we've got a fully functional iPhone-esque web app, why not turn it into a legit iPhone app you can download from the app store? With PhoneGap, it's unbelievably easy. I'm serious here. The first time I did it, I just couldn't believe I was seeing my app in the iPhone simulator. It took me less than half an hour to figure it all out. You're seriously going to be blown away by this.

What You Need to Get Started

Some setup is required. You'll need to download Apple's iPhone Software Development Kit (SDK) and PhoneGap. You'll also need to have all of your web app's files on hand. PhoneGap relies on them to build your iPhone app.

Here's a more comprehensive list:
iPhone SDK from Apple's website
(http://developer.apple.com/devcenter/ios/index.action)
PhoneGap from www.phonegap.com
A Mac laptop running the latest version of Snow Leopard
A test device (iPad, iPhone, or iPod Touch)
All of your web app's CSS and Javascript code in one place

How to install iOS SDK and Xcode

Click on the Xcode and iOS SDK link once you reach the page we linked to above. You'll need to have your Apple ID and password on hand. If you've never purchased something from iTunes, the App Store, or Apple, you'll need to create one. Here's the form where you can download the SDK.

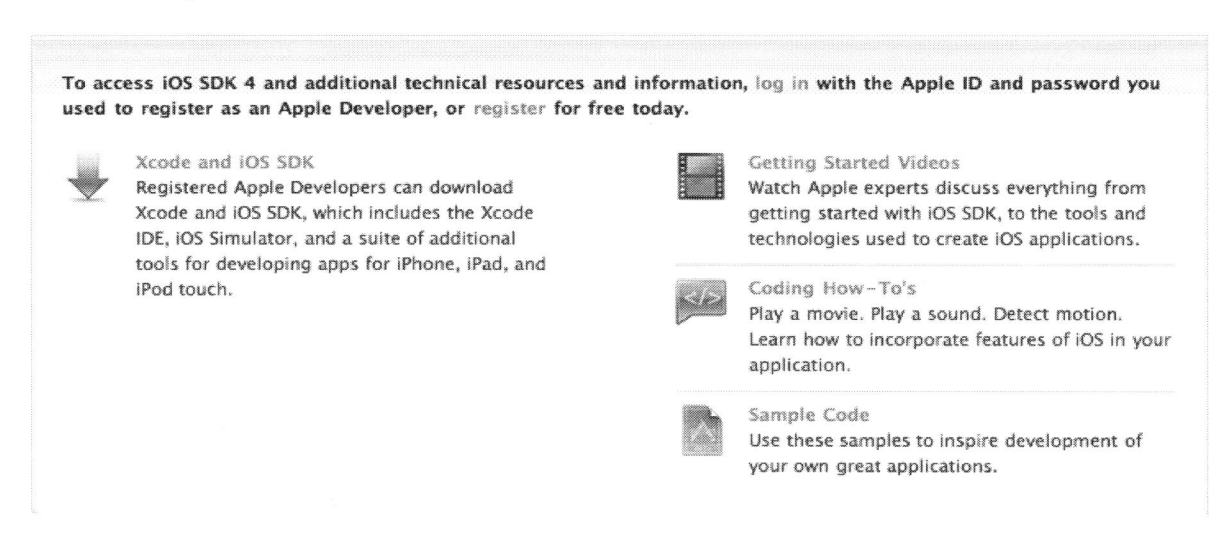

You'll eventually need to sign up for Apple's developer program at a fixed cost of $99. That's a bridge you can cross later on. For now, you'll still be able to test your app on the iPhone simulator, which for the purposes of this guide, is good enough. Also be aware that it takes about 2 hours to download and install the iPhone SDK on your laptop.

Oh, and another thing. Don't try to install PhoneGap until you've already installed your iOS SDK and Xcode. PhoneGap mostly functions as a plugin, so it simply won't install properly without them.

<h2 style="text-align:center">How to Install PhoneGap</h2>

It's really easy. Download PhoneGap from the website, and then open the PhoneGap folder from your downloads folder in the finder. Here's a screenshot.

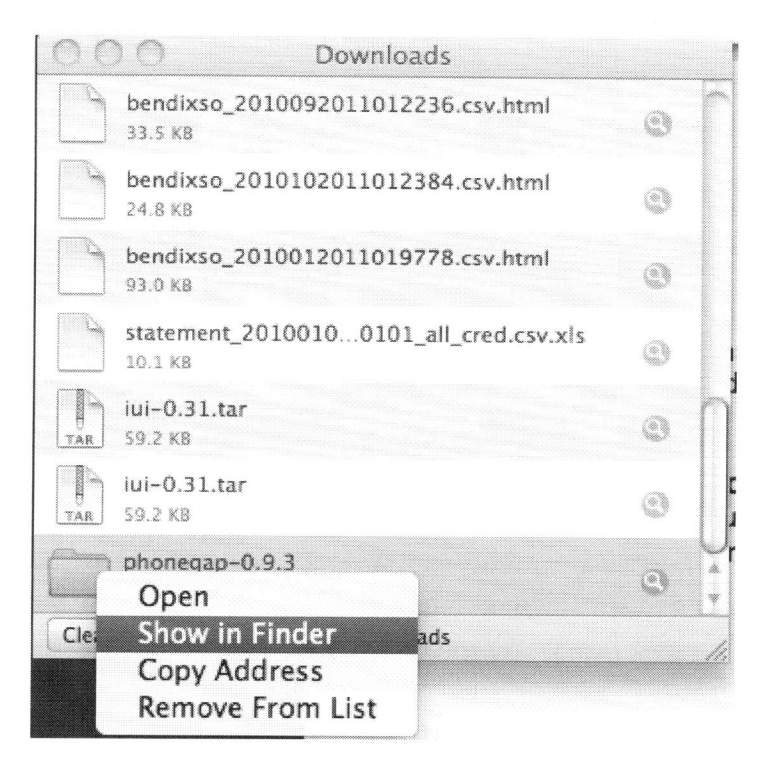

Once you've got your PhoneGap directory open, simply navigate to the folder containing the iOS install package.

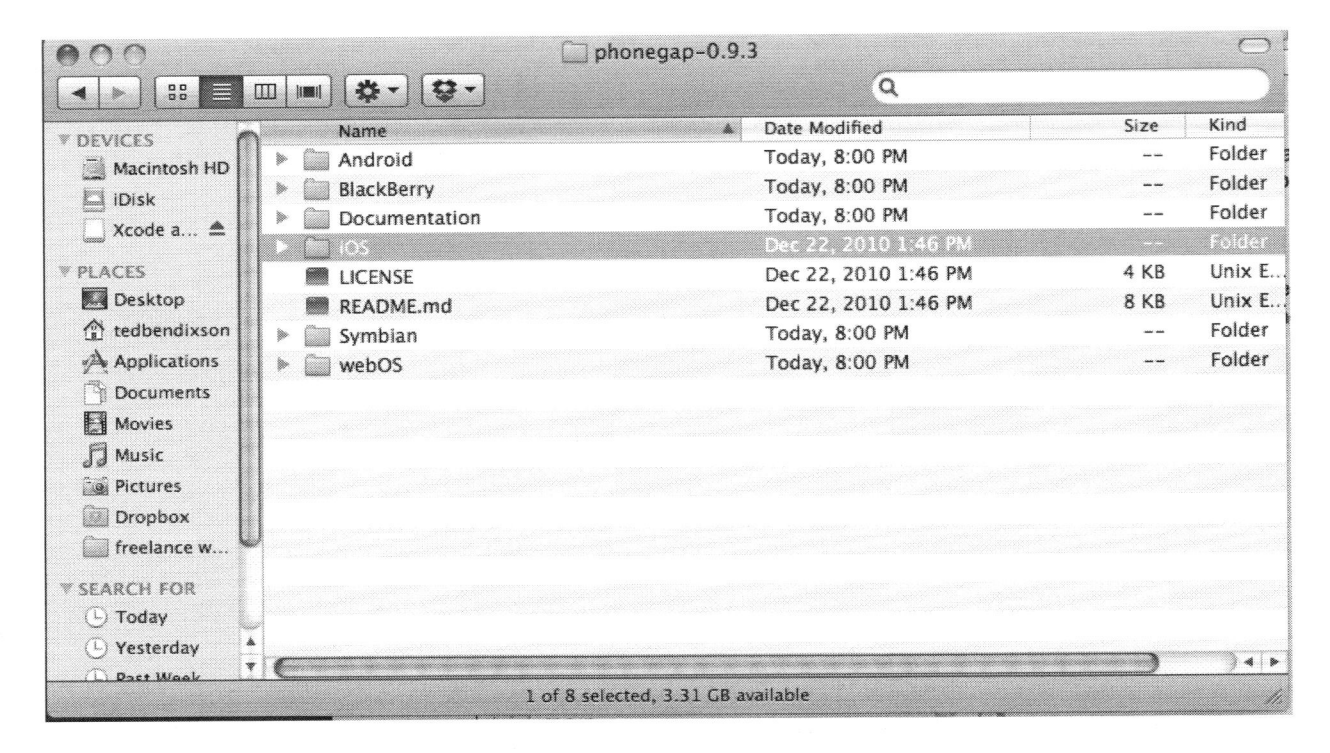

From there, the process is nearly self-explanatory. PhoneGap will install itself, locating your Xcode program all on its own. Now you just need to get started with your new PhoneGap projects.

Putting PhoneGap to Use

Open up Xcode and start a new project. You can get to Xcode by clicking on the "developer" folder from your home folder. After that, click on applications, and you should find Xcode in there.

The next screen will ask you to pick a template for your project. Under the heading "user templates," you should see PhoneGap. Select it and click choose.

Now you can create a name for your app. For the time being, we're sticking with the skiing app we built in the last chapter. I called mine "shredtastic tuesdays," but any name works fine. Just make sure you can remember it.

And with that, you'll be taken right into the completely unfamiliar world of Xcode. Don't worry about this strange and foreign place. You only need to know a few things to turn your web app into an iPhone app. Do you see the screenshot above? It says you only need to modify the contents of the "www" folder to get started.

So let's do that. Let's copy all of your web app's files to the "www" folder. I did it by double clicking on the www folder, opening it in the finder, and then copying the files over from my website folder. Make sure you get the iUI folder and all of its contents. You want to make sure you have the javascript libraries, the images, and all of the components that PhoneGap and Xcode will use to build your user interface.

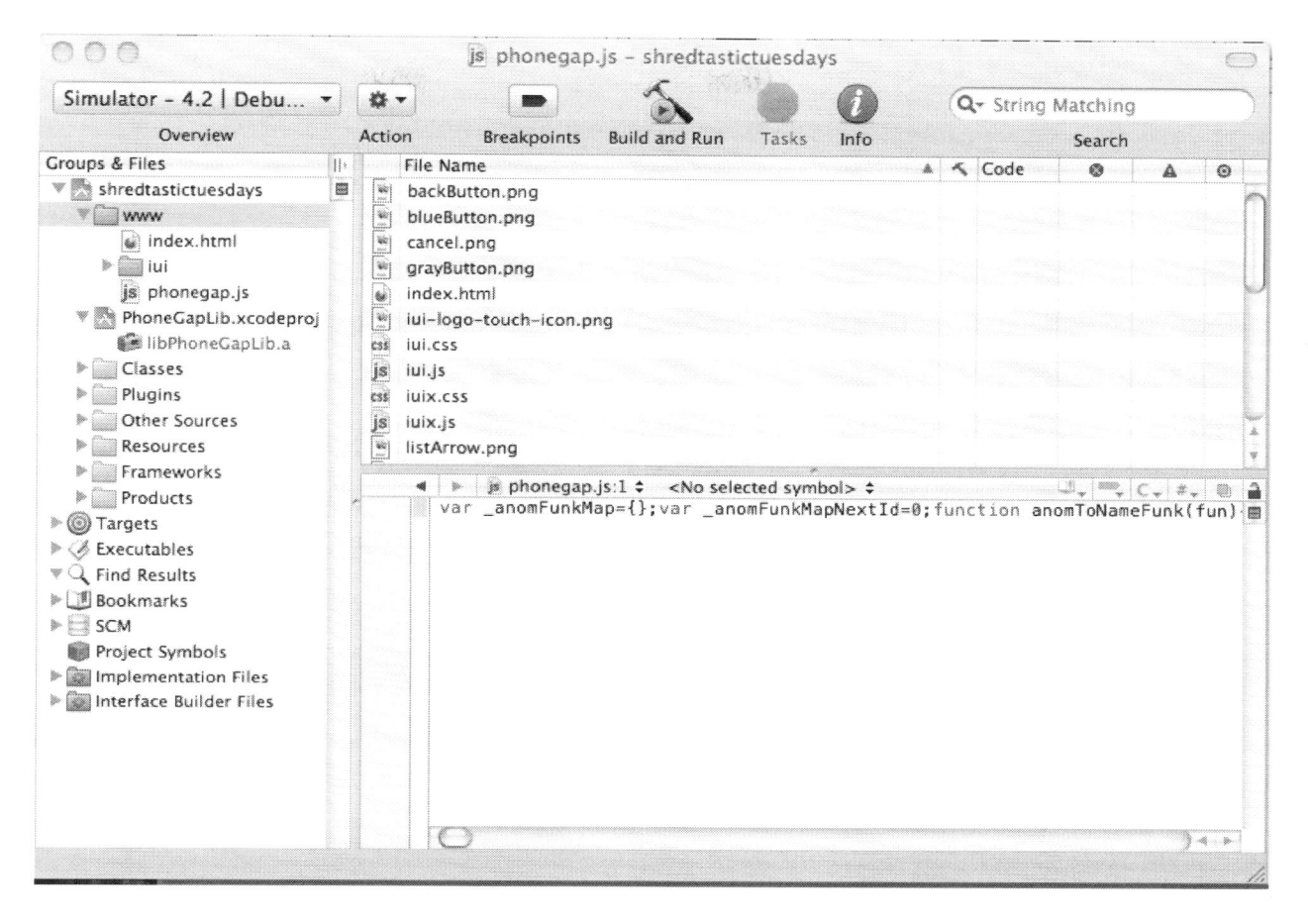

Don't Attempt to Build Your App Just Yet! Restart Xcode.

You won't find this step in the official instructions, but it's absolutely crucial. The first time I attempted to build my app, Xcode gave me a bunch of errors. Not cool. It turns out that Xcode couldn't find my PhoneGap files, making the build impossible. I restarted Xcode and everything went off without a hitch. I was looking at a simulated version of my app within minutes.

You should see the phonegap.js file in your "www" directory before you press the Build and Run button. That file is PhoneGap's way of, literally, bridging the gap between iOS and Javascript/CSS. Once it's there, you should be able to build your app without any errors.

And voila! Here it is, the exact same app on a simulated iPhone. You can't tell me that isn't absolutely mind-blowing, especially considering how little work you had to put into it.

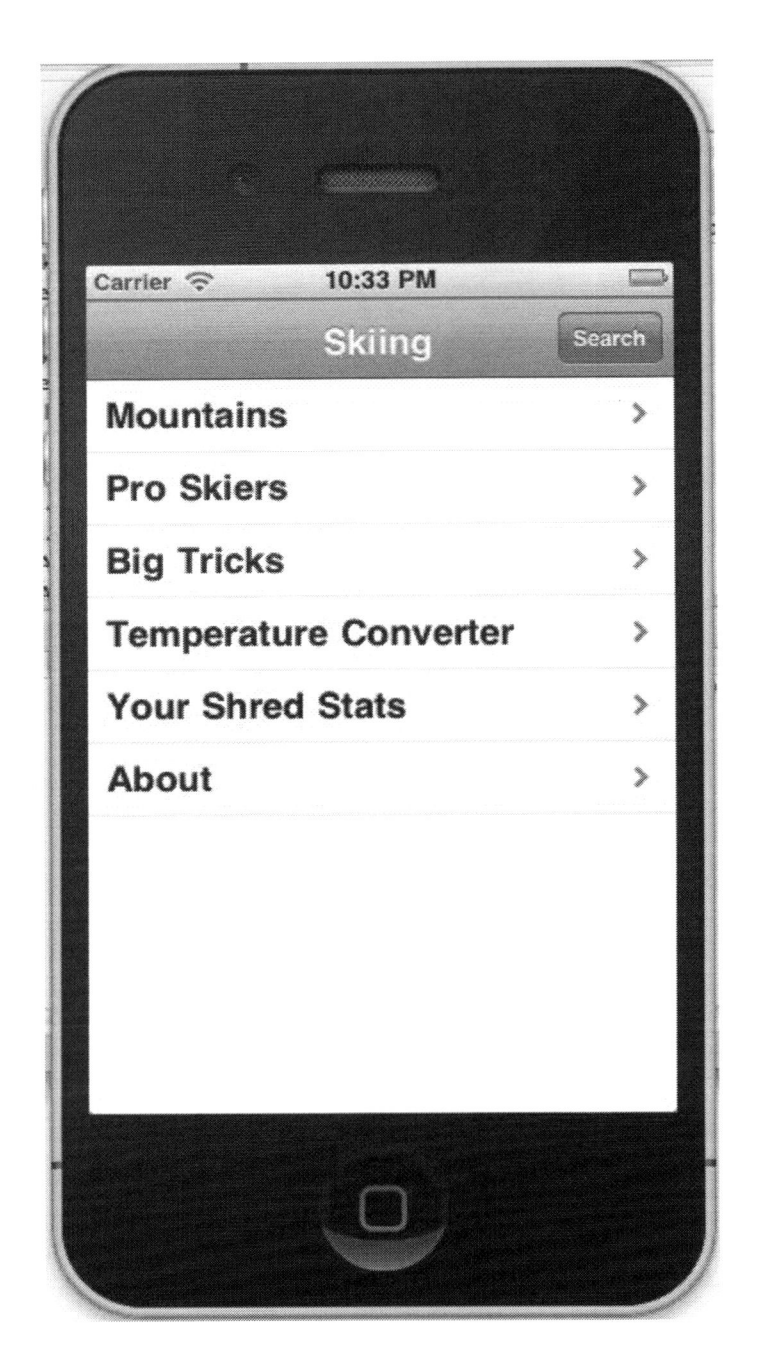

I encourage you to take some time to play around with the app. When I tested it, everything was exactly the same as the web version. Very impressive.

Now It's Time to Test, Test, Test

We've come a long way, having gone from knowing nothing about mobile design and web apps to building our own iPhone-esque app and testing it in the simulator. What's even more amazing is that we've done it all with little to no knowledge of the actual code we're using. We're rolling along smoothly on the wheels others have invented.

So what can we do from here? iUI certainly is nice, but there's something even better. It's called Sencha Touch, and it's loaded full of options. In the next section, we'll show you how to build some beautiful apps with Sencha Touch. We'll get a little more in-depth, building advanced user interfaces and slightly more complicated functionality. You seriously won't believe how easy it is to get the job done with this powerful tool.

Chapter 7: Building A Web Application with Sencha Touch

So far, what we've done is fairly basic. We've picked up a template for creating quick-and-easy iPhone apps with HTML, and we've made an iPhone-eqsue skiing app. While this is certainly an accomplishment (especially if you don't already have a web app for your business) there isn't much that we can do with iUI-based apps. If we want to customize our app to an even greater extent, we need to look at some other software development packages.

Sencha Touch is one such package. With platforms like Sencha Touch, you can create your own touch-based user interface with a few fairly simple commands. You can tell Sencha Touch where you want to place certain buttons, touch-based panels, tabs, and a bunch of other web app design elements. This gives you much more flexibility and wiggle room to create an app that better suits your needs.

Once you're done creating your user interface with Sencha Touch, you can add in some functionality with Javascript. Then it's off to PhoneGap. It's the same thing we did with iUI, except we'll be able to custom craft almost every aspect of our design. So, without any further ado, let's install Sencha Touch and see what it can do for our web apps.

How to Install Sencha Touch

Just like iUI, Sencha Touch is one big Javascript library. To get it to work, download all of the Javascript files in one big bundle from Sencha's website, upload them to your server, and then link to them in your HTML and Javascript files.

You can get the files from this link:
http://www.sencha.com

There are a few other options on their website. Download Sencha Touch by clicking on this button:

Sencha Touch
Build rich mobile web apps
for Android, iPhone & iPad.

Download

Learn more
Demos
API Docs

As I said, Sencha Touch is one giant Javascript library. Once you've downloaded it, you'll need to upload it to your server so you can start writing the code. If you're using cPanel, the best way to go about this is to upload the .zip file and then uncompress it into its own folder on your server. There are other ways, but this one doesn't require you to use a separate FTP client (which can be a pain because they don't always work for a variety of reasons).

Here's a screenshot from my cPanel account:

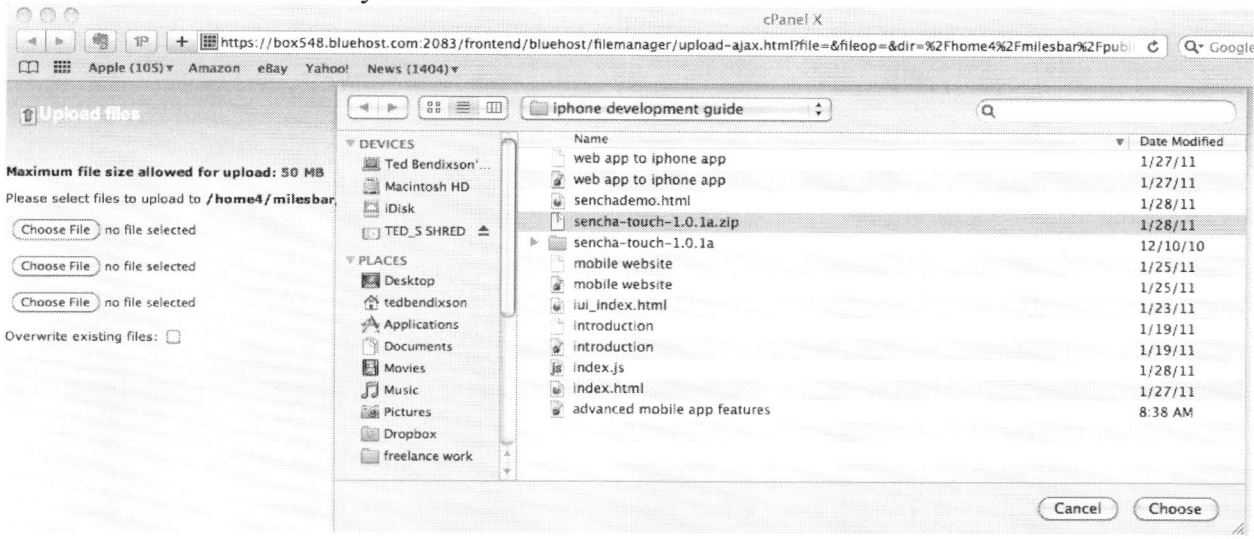

Once your package is uploaded, unpack it by clicking on this button:

Extract

Cpanel will then ask you where you want to unpack your files. I prefer to unpack mine to the main directory because Cpanel automatically creates its own senchatouch directory. I also change the name of the folder when I'm done so it's easier to link to it in my HTML and Javascript files.

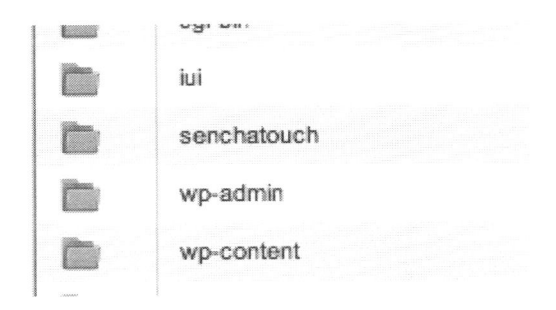

Once we've gotten this part out of the way, we need to setup our Sencha Touch development environment. That means creating the HTML and Javascript files that will contain our code.

How to Setup a Sencha Touch Development Environment

When we made our iUI app, we placed all of the HTML and Javascript in one file, the index.html file. We then did most of our coding in HTML because iUI is designed to create an entirely different user interface through basic HTML commands like <ul>, <li>, <form>, and others.

Sencha Touch works in a slightly different way. Instead of creating the user interface elements in HTML, you make them in Javascript. That means we're going to be writing a lot more Javascript to create our Sencha Touch commands. To do this, we need to separate our HTML, CSS, and Javascript files. We'll then link to all of them in our "senchademo.html" file.

We'll be using 4 files. Here's a brief summary of what each file does:
senchademo.html: This file contains the HTML that links to your CSS and Javascript files.
senchademo.js: This file will contain all of the Javascript code you will be using to build your Sencha Touch user interface and app.
sencha-touch.js: A file that contains a library of user interface elements you can create with Javascript.
sencha-touch.css: A stylesheet that defines the look and feel of the buttons and user interface elements you will be using.

Sencha Touch contains a bunch of different Javascript libraries for further customization, but for the purposes of this section, we'll only be using the ones I've discussed. You can find the sencha-touch.js file in the root directory of your Sencha Touch library, but the sencha-touch.css file is a little more tricky. It's in the resources/css directory.

How to Setup senchademo.html

This is the file we'll be using to test our Sencha Touch application. You can call this file "index.html" if you want. I'm just using this name because I didn't want to overwrite the index.html reference file that comes with the Sencha Touch software package.

Your senchademo.html file will be remarkably simple. You're simply going to link to all of your other code. Here's what it looks like.

```
1   <!DOCTYPE html>
2   <html>
3   <head>
4       <meta http-equiv="Content-Type" content="text/html; charset=utf-8">
5       <title>Layouts Buttons</title>
6       <script type="text/javascript" src="http://www.shredtastictuesdays.com/senchatouch/sencha-touch.js"></script>
7       <link rel="stylesheet" href="http://www.shredtastictuesdays.com/senchatouch/resources/css/sencha-touch.css" type="text/csss">
8       <script type="text/javascript" src="http://www.shredtastictuesdays.com/senchatouch/senchademo.js"></script>
9   </head>
10  <body></body>
11  </html>
```

None of the heavy lifting is done with the body section. It's all done in the header. Once you've created this file, upload it to your server and get started with your senchademo.js file (if you want to call this file index.js, that works too. Just remember to link to it correctly).

A Piece of Software You Should Consider Downloading

We are about to get into some pretty heavy coding. For now, a standard text editor is probably all you've needed. But programmers and web developers never use standard text editors. They use code editors that color in all of the different web design elements, functions, variables, and other commands. These are extremely handy because, when you can see the different colors, you can spot errors more quickly.

It's like having a spell checker for your code. Have a look at the screenshot above. Do you see the blue? Those are all of my HTML tags. How about the red? That's anything sandwiched in between quotes. The yellow is there to point out different attributes in the tags I've created. At a very quick glance, I can see that my code makes sense. If I were Martha Stewart, I would say this is a good thing.

There are a bunch of different code editing programs on the market. I've gone with Smultron because it's cheap, simple, and easy to use. It also has another really cool feature. Whenever I save a file with a certain extension (.js for example), Smultron recognizes which programming language I'm using. Then Smultron changes the colors based on the programming language, making it easier for me to spot errors in my code.

You can get Smultron on the Mac App Store for $4.99. That's a pretty small investment for an app that will come in handy no matter what programming project you're working on. Believe me, it will save you hours of frustration.

You'll Also Want to Do This...

Just like any programming project, you are going to make mistakes. Your code will zig where it's supposed to zag, and you'll accidentally type a comma where you should have a placed a semicolon. You can sit there for hours and hours trying to find your mistakes, or you can enlist the help of a Javascript debugger.

Most browsers have a built-in Javascript debugger. I'm currently using Safari, the browser that comes with my Mac, and I found mine under Develop --> Start Debugging Javascript.

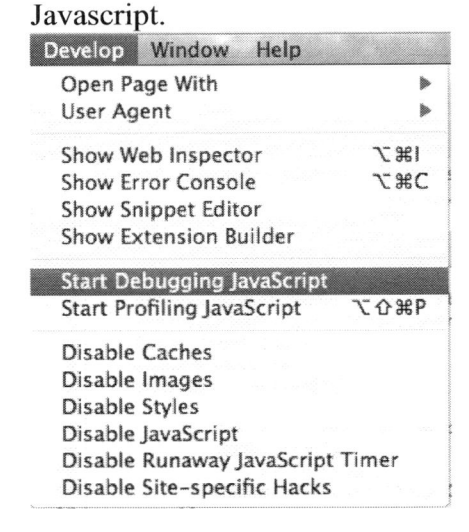

When you click on this menu item, a window with a bunch of different options will attach itself to your browser window. For the time being, click on "console," and then click on the "errors" menu item.

When you do this, you'll get a list of Javascript errors and the lines they occur on. Here's an example to start you off.

I made a mistake and forgot to put a comma after the line reading style: "background-color: #f00;" Take a look at this screenshot.

```
15          fullscreen: true,
16          style: "background-color: #f00;"
17          items: [],
```

Now, when I try to run my Javascript code, nothing happens. I just get a blank white page. Thankfully, I've got my Javascript debug console open. It tells me that it found a parse error on line 17 of senchademo.js file.

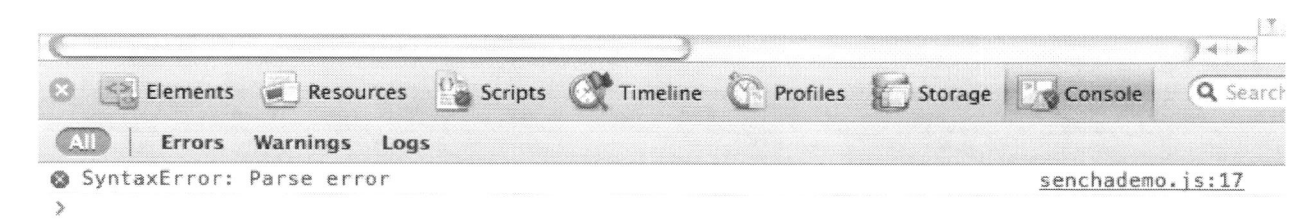

If you remember, the actual error occurs on line 16. We forgot to put a comma at the end. But, for the purposes of debugging, line 17 is close enough. We can investigate all of the code near it, and we should find the mistake much faster.

Understand that no debugging program is perfect. You will sometimes get taken on a wild goose chase through the vast code library that is Sencha Touch. In that case, you won't have much of a choice but to go through every line with a magnifying lens, trying to spot those little mistakes that can make your app crash.

Debugging programs help, but they're not a panacea for your coding problems. It also helps to have a clean and uncomplicated coding style. Always devote one line of code to one command, and avoid placing functions inside of functions inside of functions. This makes your code very difficult to read, and therefore, very difficult to debug.

Your First Sencha Touch User Interface

Now we're going to create the Javascript file that specifies all the details of your user interface. We'll call it "senchademo.js." You should recognize some of the commands from our previous chapter on Javascript, but there are many others to take note of. Many of these commands refer to some very specific Sencha Touch settings. You can only learn them by trying them out for yourself.

We're going start by specifying the Ext.setup() function. This is a Javascript function that Sencha Touch uses to setup your user interface. There are a bunch of other functions and commands inside of it, and we'll be using those to make the buttons, forms, and other design elements for your web app.

Here's what the beginning should look like:

```
1  Ext.setup({
2
3
4  });
5
```

Ext.setup() is not a standard Javascript function in the way you probably understand it. First of all, what's with the extra { } brackets? Well, those are there because Ext.setup() is actually an array that contains a bunch of different settings and setup functions.

Wait wait wait... what's an array? Think of it as a list of things. Your grocery list could be an array. If you wanted to, you could write it in the language of computer science. Here's what it would look like:

```
1  {
2       potatoes,
3       chicken,
4       oatmeal,
5       beer,
6       broccoli,
7       carrots,
8       ground beef
9  }
```

It's not all that different from your Ext.setup() function, except your Ext.setup() function contains a list of design elements along with their attributes. Here are a few of the elements you'll want to begin with.

```
1  Ext.setup({
2       tabletStartupScreen: 'tablet_startup.png',
3       phoneStartupScreen: 'phone_startup.png',
4       icon: 'icon.png',
5       glossOnIcon: false,
```

So, what are these? Well, they're links to the images your app will use for the startup screen on both tablet and phone devices. The attribute is a link to an image that will become your app's icon in the app store. And finally, the glossOnIcon is a simple true or false switch that you can set to put an extra shine on your app's icon.

But Ext.setup() doesn't just contain a list of design elements. It also contains an important function that you'll use to tell Sencha Touch to begin building your interface. We call this the "onReady" function.

Do you remember our temperature conversion program that we wrote in Javascript for our Skiing app? Do you remember how we used a piece of code called "onclick" for the button? If you don't remember, here's that code again.

```
      </div>
   </fieldset>
   <a class="whiteButton" type="submit" onclick="ctemp.value=tempconvert(ftemp.value)" href="#temperature.ctemp">Convert</a>
</form>
```

"onclick" is a button attribute that tells Javascript to run a particular function when you click on the button. "onReady" is similar in that it tells Javascript to run a function. But instead of running the function when somebody clicks on a button, Javascript runs the function when SenchaTouch is ready. Here's what it looks like in your senchademo.js file.

```
5   glossOnIcon: false,
6   onReady: function(){
7
8       }
9
```

SenchaTouch, like any program, has to go through a few steps before it is ready to build your user interface. If you tell Javascript to start running a bunch of code, and SenchaTouch isn't ready, the functions in that code might not be loaded or available to Javascript. In basic English, that means your program won't run because Javascript won't know where to find the code. So you have to wait for all of your Sencha Touch code to get loaded before you can use it to build your interface.

onReady only runs when SenchaTouch has given Javascript the signal to go ahead and build your user interface. Whatever we choose to place in this function determines how our app will start out. It's kind of like an initialization program where you set all of the beginning variables and other elements.

To get started, we're going to build a simple horizontal array of three buttons. That's it. Three buttons. We'll then build off this and go into some more advanced Sencha Touch commands to give you an idea of where the possibilities lie. Considering how complex this is starting to get, I think you're pretty well aware that you can do almost anything. That's the most exciting thing about this particular Javascript library. Endless design possibilities!

A Simple Three Button Interface in Sencha Touch

I rarely do this, but I think it makes sense for this particular example. I'm just going to paste the code right here, and then I'll go into a more in depth explanation. These are the commands we're going to use to build a simple three-button interface.

```
onReady: function(){
    new Ext.Panel({
        fullscreen: true,
        layout: {
            type: 'hbox',
            pack: 'center',
            align: 'center',
        },
        defaults: {xtype: 'button'},
        items: [
            { text: "button 1" },
            { text: "button 2" },
            { text: "button 3" }
            ]
    })
}
```

We already discussed the onReady() function and what it does. But what is this Ext.Panel() thing? To put it simply, an Ext.panel() is a design element that comes with Sencha Touch. When you use the "new" command, you're creating one of them in your user interface. Just like Ext.setup(), an Ext.panel() is a list of things and their attributes.

An Ext.panel() can be fullscreen or non-fullscreen. We have it set to "true," so it takes up the entire screen, regardless of the device we're using. But if you set it to "false," it will only take up part of the screen. This is just one of many different settings you can tweak in the Ext.panel() list.

By now, you've probably noticed that your design will work by placing lists inside of lists and functions inside of functions. That's okay, and it's actually a much more elegant way to build your app. In this demo, we're placing a few buttons inside of a panel. But you could place a panel inside of a panel and then some buttons inside of that. It's all about the design you have in mind. The code is there to help you get there.

So let's go over a few more attributes in the Ext.Panel() function. Here's a list:

Layout: This is a list of attributes that determine the way the panel is arranged on the screen.
Layout type: You'll see that we've chosen 'hbox.' This represents a horizontal box with elements that go across the screen. You'll see this when we add our buttons into the design.
The 'pack' attribute: This determines how the elements will fill up your panel. When you pick 'center,' they'll be in the center of the panel. When you pick 'start,' they will appear at the beginning of the panel. You can also pick 'end,' a setting that places the buttons at the end of the panel.

The 'align' attribute: This attribute has two options. You can pick 'center,' which places the elements in the center of the panel, or you can pick 'stretch,' an option that stretches the elements to fill the panel.

Defaults and xtype: This is where you tell Sencha Touch what's going inside of your panel. In this case, we're putting buttons in the panel. But it could be anything. We'll discuss more of this later.

The items array: Here, we're just telling Sencha Touch that we want to create three buttons, each with their own text on them.

Considering the way we've defined our attributes, what do you expect to see once we run our Javascript program? How will Sencha Touch display the buttons?

Here's a screenshot of what happens when you run this code:

We said above that we wanted a horizontal panel with buttons packed and aligned in the center. That's exactly what we got. Pretty cool. What else can we do?

Now let's modify our code a little bit so we can change the orientation of our buttons. Here's the entire code with our slight modifications. This is a good time to make sure you've closed off all the functions and arrays with the proper brackets and end parentheses. If you've got a text editor like Smultron, this is much easier because it will find the opening bracket and show you which closing bracket corresponds to it. That way, you can be certain each ending bracket has a matching beginning bracket.

```
 1  Ext.setup({
 2      tabletStartupScreen: 'tablet_startup.png',
 3      phoneStartupScreen: 'phone_startup.png',
 4      icon: 'icon.png',
 5      glossOnIcon: false,
 6      onReady: function(){
 7          new Ext.Panel({
 8              fullscreen: true,
 9              layout: {
10                  type: 'vbox',
11                  pack: 'start',
12                  align: 'center',
13              },
14              defaults: {xtype: 'button'},
15              items: [
16                  { text: "button 1" },
17                  { text: "button 2" },
18                  { text: "button 3" }
19                  ]
20          })
21      }
22  });
23
24
```

Now we've changed the type to "vbox," a command that specifies a vertical box for the application. We've also changed the pack attribute to "start" instead of center. Here's what we get when do that.

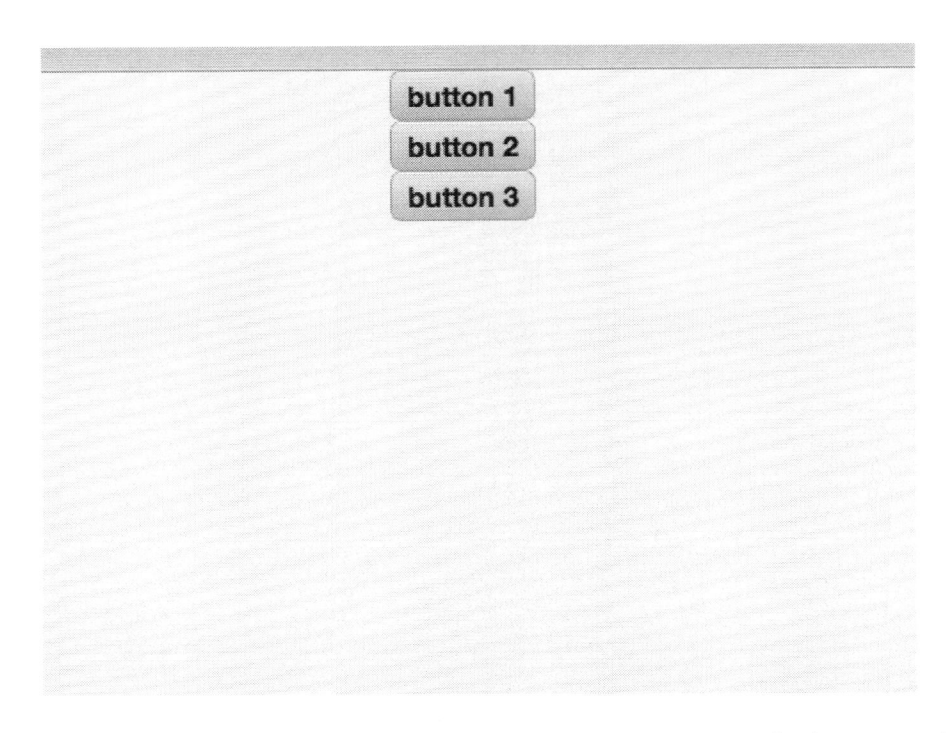

Just as we could have predicted, the buttons are now vertical, center-aligned, and they appear at the beginning of the vertical box panel we've created.

While buttons are certainly a cool thing to have, we can't really do anything with them unless we've got some sort of area we can modify whenever someone presses our button. That's why the next section will focus on all the different panels you can use in Sencha Touch, how to arrange them, and ultimately how to use them to build your web app.

How to Use Panels in Sencha Touch

Websites, apps, and windows all have panels. They're full of spaces where you put things. Most of your designs in Sencha Touch will be built from panels that you place next to one another, inside another, and in any configuration of your choice. Sencha Touch is awesome because, just like iUI, it comes packed with a bunch of useful panels that practically build your user interface for you.

But before we start to use some of these more specialized panels (for example a tabbed panel that allows users to switch between different pages), there are a few things you should learn about panels in general. In this short tutorial, I'll show you some commands you can use to arrange your panels in any configuration you can imagine.

So, let's go back to the drawing board. We have already created one panel, the Ext.panel(), but we can't actually see it because it is white by default. However, if we change its background color, you'll be able to see that the panel really does take up the entire screen.

Here's the code that gets us back to the drawing board:

```
1   Ext.setup({
2       tabletStartupScreen: 'tablet_startup.png',
3       phoneStartupScreen: 'phone_startup.png',
4       icon: 'icon.png',
5       glossOnIcon: false,
6       onReady: function(){
7           new Ext.Panel({
8               style: "background-color: #f00;",
9               fullscreen: true,
10          })
11      }
12  });
13
14
```

You'll see that I've added a background color to make the panel more visible. If you know your hex, that value corresponds to the color "red." Here's what we get when we run the code on our server.

Who could've predicted that? It's a big blob of red, and it takes up the whole screen. It doesn't do much, but you'll just have to trust me that we're onto something here.

Every panel, even the more advanced panels we'll be using later on, contains two lists of items. One of the lists, called the "items array," contains whichever items we want to display in our panel. The second list, called the "docked items array," contains items we want to dock to the top, left, right, or bottom of our panel. If you're familiar with using a Mac (which you most assuredly are), the docked items act like the dock with all the shortcuts to your favorite programs.

Here's a look at the empty code with no items in either the items or docked items arrays.

```
1   Ext.setup({
2       tabletStartupScreen: 'tablet_startup.png',
3       phoneStartupScreen: 'phone_startup.png',
4       icon: 'icon.png',
5       glossOnIcon: false,
6       onReady: function(){
7           new Ext.Panel({
8               style: "background-color: #f00;",
9               fullscreen: true,
10              items: [],
11              dockedItems: []
12          })
13      }
14  });
15
16
```

So, what happens if we add an item to the dockedItems[ ] array? Let's try it out. Let's dock some html to the bottom of our panel.

```
1   Ext.setup({
2       tabletStartupScreen: 'tablet_startup.png',
3       phoneStartupScreen: 'phone_startup.png',
4       icon: 'icon.png',
5       glossOnIcon: false,
6       onReady: function(){
7           new Ext.Panel({
8               style: "background-color: #f00;",
9               fullscreen: true,
10              items: [],
11              dockedItems: [
12                  {
13                      dock: "bottom",
14                      html: "This is docked to the bottom, and it's blue",
15                      style: "background-color: #00f;"
16                  }
17              ]
18          })
19      }
20  });
21
22
```

You'll see that I've set the "dock" attribute to "bottom," I've added some html, and I've set the background to the color blue. Check out what happens when I run this code on the server.

This is docked to the bottom, and it's blue

Great. We've got two panels, and one of them is docked to the bottom of the page. But what happens when we add more items to the dockedItems [ ] array? What will Sencha Touch do?

Let's find out by adding another item to the dockedItems [ ] array. This time, we'll dock it to the right. Here's the code for that:

```
1   Ext.setup({
2       tabletStartupScreen: 'tablet_startup.png',
3       phoneStartupScreen: 'phone_startup.png',
4       icon: 'icon.png',
5       glossOnIcon: false,
6       onReady: function(){
7           new Ext.Panel({
8               style: "background-color: #f00;",
9               fullscreen: true,
10              items: [],
11              dockedItems: [
12                  {
13                      dock: "bottom",
14                      html: "This is docked to the bottom, and it's blue",
15                      style: "background-color: #00f;"
16                  },
17                  {
18                      dock: "right",
19                      html: "This panel is docked to the right, and it's green",
20                      style: "background-color: #0f0;"
21                  }
22              ]
23          })
24      }
25  });
26
27
```

And here's what our masterpiece looks like when we upload it to the server. You can see that docking depends on the order of the items in the dockedItems[ ] array. Because the blue panel is first in the array, it gets docked to the bottom of the red panel first. Only after it is docked, can the green panel get docked to the right.

If we were to switch the ordering around, the green panel would take up the entire right side, and the blue panel would only take up part of the bottom. It would look like this:

Now, where might something like docking become really useful? Do you remember our tutorial on iUI from the other chapter? Our user interface had a toolbar at the top with buttons our user could press to navigate through the app. Well, we can create the same toolbar in our app by using the dockedItems [ ] array. Here's how.

How to Use the DockedItems [ ] Array to Create a Toolbar.

Sencha Touch provides us with a bunch of different panel types. One of them is the toolbar panel, a toolbar that draws its inspiration from the iOS user interface. You can create a toolbar much in the same way you would create a panel. You tell Sencha Touch a command like the following.

```
1  new Ext.Toolbar({
2      layout: { pack: 'center' },
3      defaults { iconMask: true, ui: 'plain' },
4      items: [ {
5
6      }]
7  });
```

Notice that this isn't all that different from creating a new Ext.panel(). That's because panels and toolbars, in Sencha Touch, belong to the same family of user interface elements. They both respond to the same layout commands and defaults. They also both have an items: [ ] and dockedItems [ ] array.

So, how might we make a toolbar for our big red panel? We're going to do it by adding a new Ext.Toolbar to our panel's dockedItems [ ] array. We first create the toolbar as a variable, and then we add it to the dockedItems [ ] array. This is a much more clean way to create the toolbar because it avoids a bunch of complicated nesting of panels and toolbars. Here's what our new code looks like.

Live Find

```
 1 Ext.setup({
 2     tabletStartupScreen: 'tablet_startup.png',
 3     phoneStartupScreen: 'phone_startup.png',
 4     icon: 'icon.png',
 5     glossOnIcon: false,
 6     onReady: function(){
 7
 8         var toolbar = new Ext.Toolbar({
 9             dock: 'top',
10             title: 'Skiing',
11             items: [ {text: 'Temp Converter'}]
12         });
13
14         new Ext.Panel({
15             style: "background-color: #f00;",
16             fullscreen: true,
17             items: [],
18             dockedItems: [toolbar]
19         });
20     }
21 });
22
```

Each item in a toolbar corresponds to a button. We can add as many as we want, but for now, we'll just create one. You'll also notice that we can specify a title. In this case, we chose "Skiing," just like our other app. But you don't have to use a title. You can go without one if you want to leave more space for your buttons. Here's what this code produces.

Aha. Everything is starting to look familiar once more. There actually is a way to completely re-create the skiing app we built with iUI using Sencha Touch, but I'd rather spare you the details. We already did that.

A Brief Respite and a Handy Resource

Throughout the course of developing my own mobile apps with Sencha Touch, I have found it extremely useful to look at the proper usage for every different panel type. You can find this in the Sencha Touch API documentation on Sencha's website. Here's a link to that.

http://dev.sencha.com/deploy/touch/docs/

All of the panel types can be found under the "Ext" menu. There are a few folders up top, and the types are below.

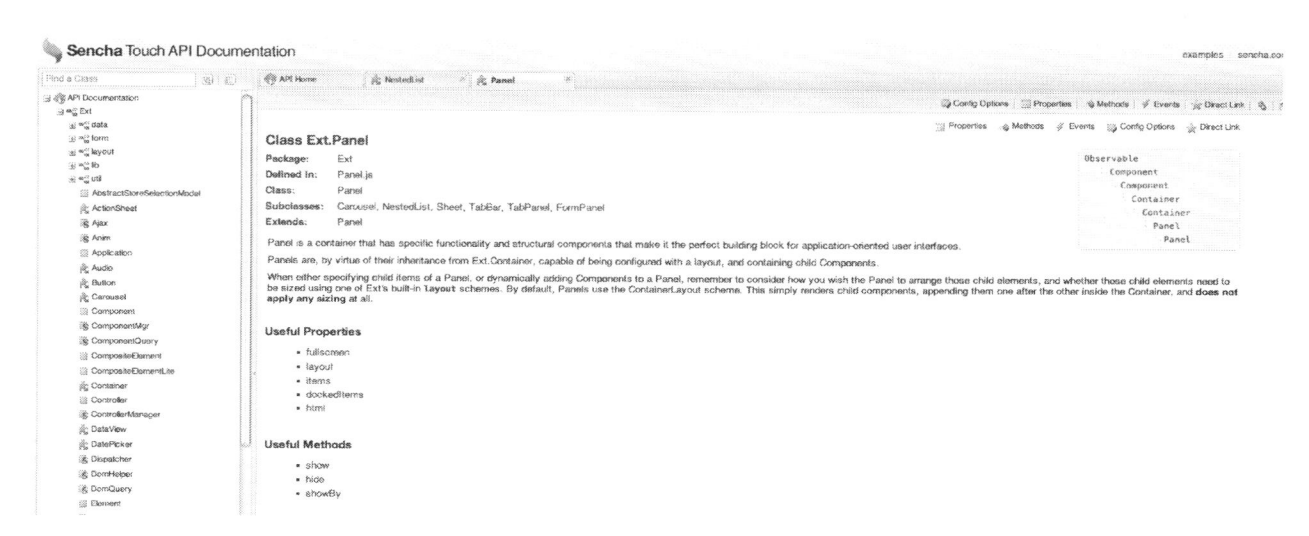

There's a reason I'm telling you this. You can't search for any of Sencha's panel types in Google. If you type "Ext.panel" into Google, you'll just get a bunch of spam pages with useless links to more spam pages (one of them wouldn't let me leave. Very reminiscent of 1997). There's a laundry list of reasons why this is so, but you don't need to know that. I'm telling you so you can avoid wasting the 2.3 hours I spent trying to find the Ext.toolbar, Ext.NestedList, and bunch of other panel types. Seriously, everything you need is on this page.

The TabPanel User Interface Element

Now we're going to try out one of Sencha's pre-made user interfaces. It's called a Tabpanel, and it works much like many other interfaces you've seen before. There is a list of buttons at the top, and a panel that switches when you tap them. It's basically an animated version of the top menu bar that's found on most websites.

The code to create a TabPanel is very similar to code you use to create a regular panel (part of the reason why I showed you this earlier). There are a few extra settings and attributes that you need to pay attention to. Each individual tab, for example, needs to have a title. I've given each one a different color, just so you can see the difference between panels as we interact with it.

Here's the code you can use to create a TabPanel. It's also available in the online Sencha Touch API Documentation along with every other panel type.

```
 1  Ext.setup({
 2      tabletStartupScreen: 'tablet_startup.png',
 3      phoneStartupScreen: 'phone_startup.png',
 4      icon: 'icon.png',
 5      glossOnIcon: false,
 6      onReady: function(){
 7
 8          new Ext.TabPanel({
 9              fullscreen: true,
10              ui: 'dark',
11              sortable: true,
12              items: [
13              {
14                  title: 'Mountains',
15                  html: 'Mountains to ride',
16                  style: "background-color: #f00"
17              },
18              {
19                  title: 'Temperature Converter',
20                  html: 'A temperature converter',
21                  style: "background-color: #00f"
22              },
23              {
24                  title: 'About us',
25                  html: 'Read more about us',
26                  style: "background-color: #0f0"
27              }
28              ]
29          });
30
31      }
32  });
```

Note that each pane can contain an entire HTML page. I haven't done it here, but you can also link to some CSS. Even if you don't want to go any further with Sencha Touch, you can at least create a really slick looking toolbar for your website.

When we run the code on our server, we get something that looks like this:

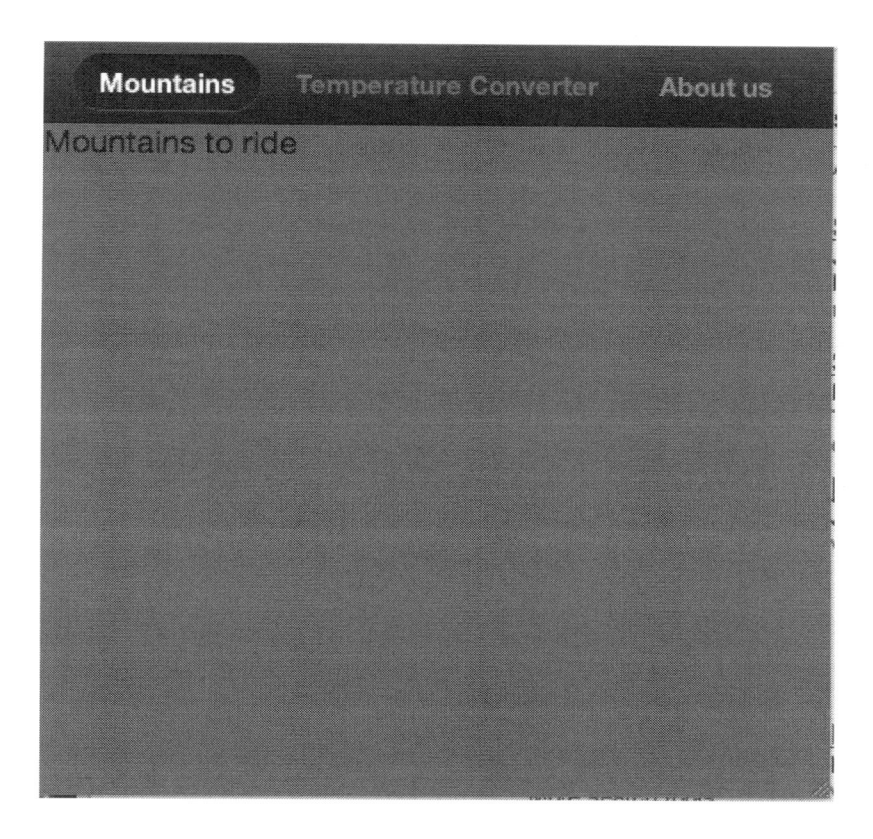

And then, when we click on the Temperature Converter, the red panel slides to the left to reveal a blue panel.

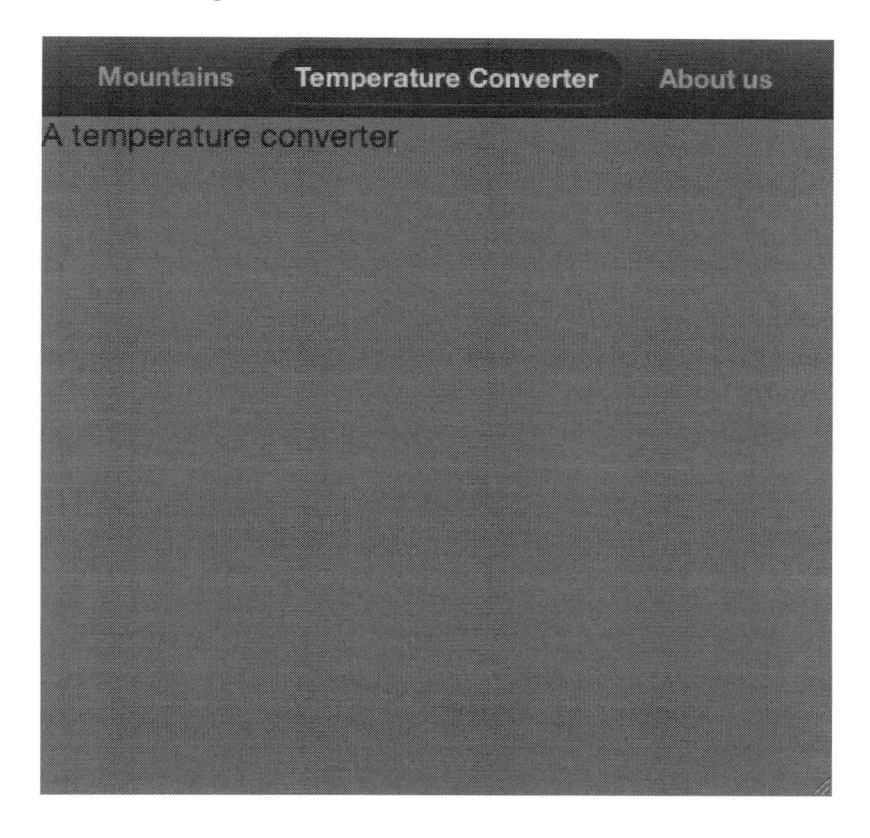

But we haven't stopped there. We can tell Sencha Touch to use a different transition animation whenever we click on one of the buttons. There are a bunch of transitions we can use including fade, slide, flip, cube, pop, and wipe. To change our transition from the default slide transition, we use the cardSwitchAnimation attribute. Here's what it looks like in code.

```
new Ext.TabPanel({
    fullscreen: true,
    ui: 'dark',
    sortable: true,
    cardSwitchAnimation: 'flip',
    items: [
    {
        title: 'Mountains',
        html: 'Mountains to ride',
        style: "background-color: #f00"
    },
```

I obviously can't show you what this transition looks like here. You'll have to try it for yourself. In any case, it's fun to watch, and your users will definitely be impressed.

If you remember from earlier, I said that you can dock more panels to any panel. Well, what if you want to add your company name at the bottom of every page on your app? With this design, you can do it all with one line of code. You simply add a panel to your TabPanel's dockedItems[ ] array.

```
items: [
{
    title: 'Mountains',
    html: 'Mountains to ride',
    style: "background-color: #f00"
},
{
    title: 'Temperature Converter',
    html: 'A temperature converter',
    style: "background-color: #00f"
},
{
    title: 'About us',
    html: 'Read more about us',
    style: "background-color: #0f0"
}
],
dockedItems: [
{
    dock: 'bottom',
    html: 'Produced by Shredtastic Tuesdays'
}
]
```

Because the bottom panel is completely separate from the top TabPanel, it doesn't change when you click on the buttons up top. Here's a screenshot after I clicked on the temperature converter button.

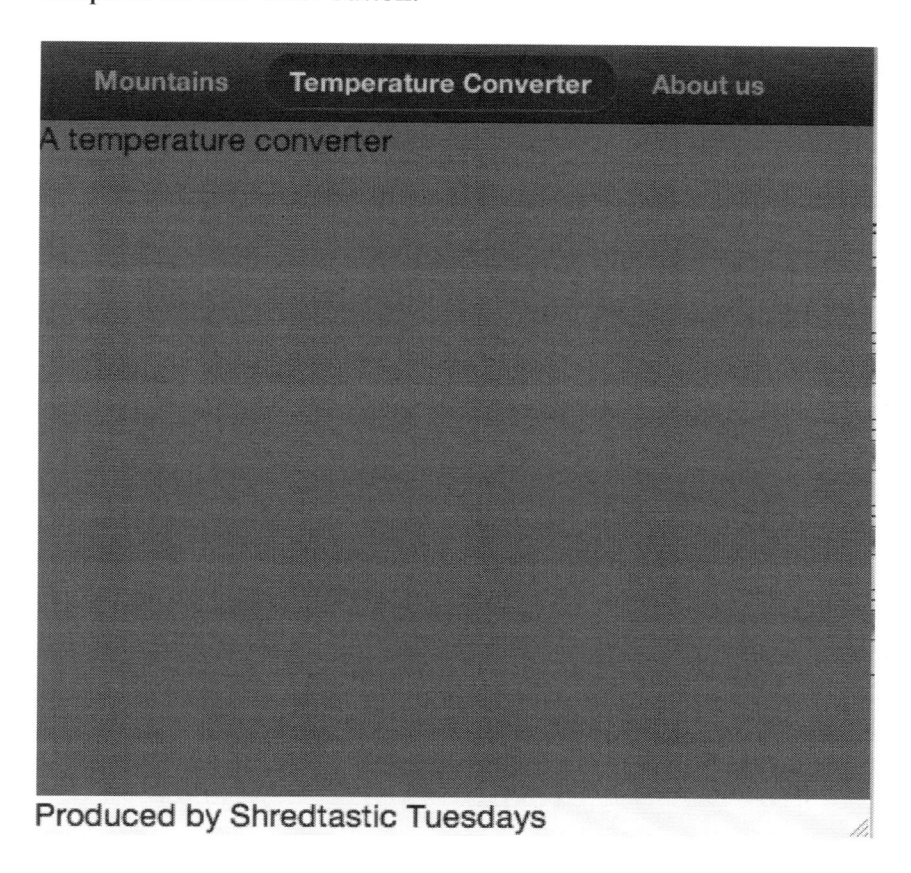

If you were to do this in a more stylish way, you would make the bottom panel match the top panel. I'm just keeping it as is so you can see where the panel is positioned.

Because Sencha Touch is built for mobile devices like the iPhone and the iPad, this interface will look the same even if you tilt the iPad on its side. That's the main reason why we've decided to dock panels in this way. With Sencha Touch, it doesn't matter how your user holds the device. Everything is automatically corrected for you.

Now that we're starting to get the swing of things, let's add some actual functionality to our app. How about an embedded Google map? You'll be surprised with how easy it is to add one. Let's give it a try.

Let's Add a Map to Our App!

Before we can use Google maps in our app, we need to link to the Google maps API. This isn't that big of a deal. We simply add one extra line of code to our senchademo.html file. Here's what it looks like:

```
<!-- The following line of code links to the Google Maps API -->
<script type="text/javascript" src="http://maps.google.com/maps/api/js?sensor=true"></script>
```

You have to place this line of code between the <head> tags in your sechademo.html file. The order isn't particularly important. It can go anywhere near the other links to your javascript files.

Next, you need to create a map object. The process is very similar to what we've been doing all along. We declare a map variable, and then we assign a few properties to it. Once we've done that, we'll add it to the items: [ ] array in our TabPanel.

Here's the code we'll be using to create a map:
```
onReady: function(){

    var map = new Ext.Map({
        title: 'Ski Map',
        getLocation: true,
    });

    new Ext.TabPanel({
        fullscreen: true,
```

For the time being, we've given it some pretty basic information. The rest is up to Sencha Touch. Notice that we have to create the map *before* we create the TabPanel. If we were to do it the other way around, our app wouldn't work because Sencha wouldn't know where to find the map that we're going to embed in the TabPanel.

And this brings me to my next point. How do we get the map into the TabPanel? It's pretty easy. We just plop it into the items: [ ] array and Sencha Touch takes care of the rest. Here's the code we'll use to do that:

```
new Ext.TabPanel({
    fullscreen: true,
    ui: 'dark',
    sortable: true,
    cardSwitchAnimation: 'flip',
    items: [map,
    {
        title: 'Mountains',
        html: 'Mountains to ride',
        style: "background-color: #f00"
    },
```

Did you catch that? It's the first thing we're inserting into the items: [ ] array. Because we called the map "map," we just write "map" in as the first item. You don't need to surround it in curly braces { } or anything like that. Just write "map" and add a comma.

Here's what you get when you run the code. Suddenly, the first item of our TabPanel has been replaced by "Ski Map," the title of our map (look above if you don't believe me).

Because it is the first item, our map is the first thing we see when we open our app. Very cool.

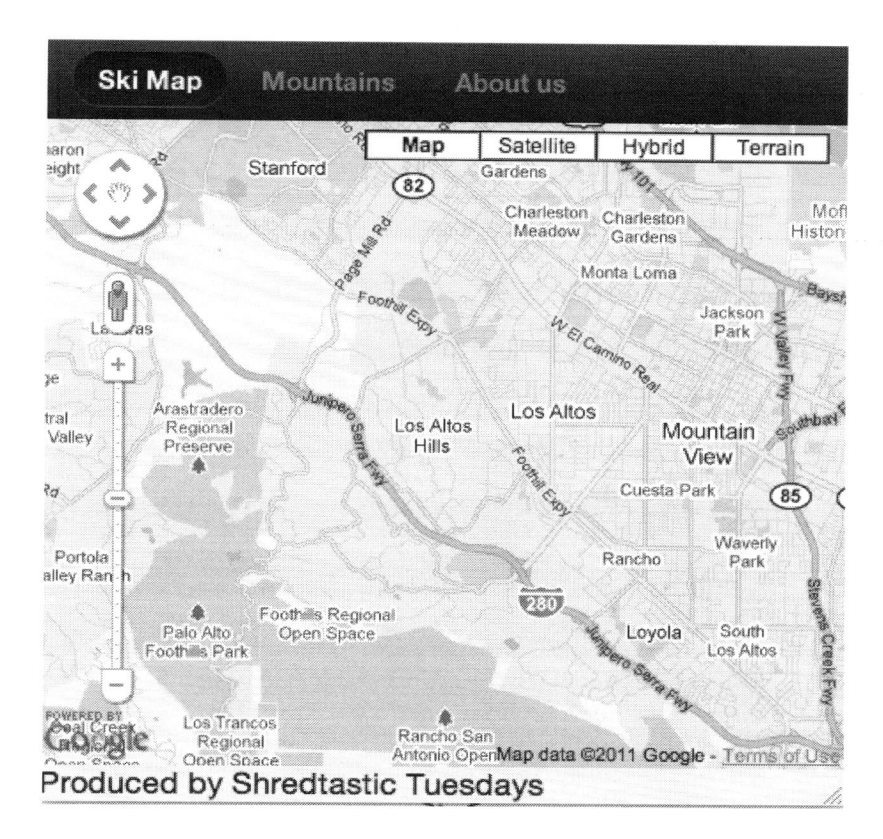

This is all well and good, but I don't live in Los Altos California. What if I want to change the starting location to something more appropriate, say Breckenridge Colorado? How might we do that?

We just have to add an extra command to our OnReady() function, right after we create our TabPanel. The command we're using is a sub-command that's available from the map we just made. It's called the map.update() command. We only need to feed it some coordinates to get started.

Here's the code you can use to change your map's starting location:

```
map.update(
{
    latitude: 39.51,
    longitude: -106.06
}
);
```

According to the National Weather Service, these coordinates place us right in the middle of Breckenridge. When we refresh our app, we get this image as the starting map.

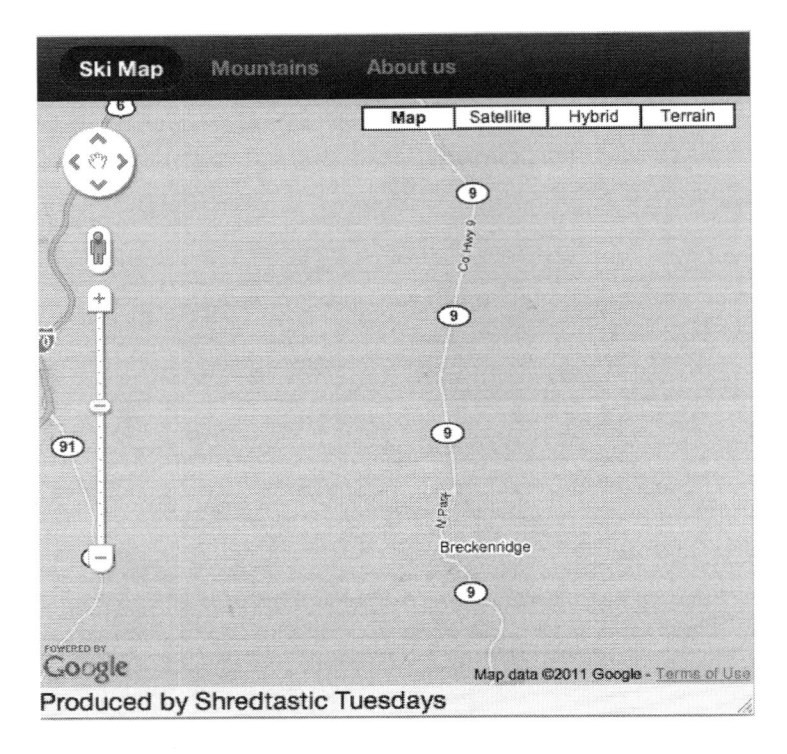

Produced by Shredtastic Tuesdays

This is kind of nice. At the very least, you can see where Breckenridge is located. But there really isn't that much information present. Maybe it would be better if we zoomed in a little more so we could see more of the city. No problem, we just change our mapOptions and pick a zoom level we prefer.

```
var map = new Ext.Map({
    title: 'Ski Map',
    mapOptions: {
        zoom: 15
    }
});
```

Now, when we refresh our map, we should get a clearer picture of the different roads near Breckenridge. Remember, we're using the National Weather Service's coordinates, so they aren't exactly tied to the center of town. These are the coordinates where they take their weather readings. Here's our new map.

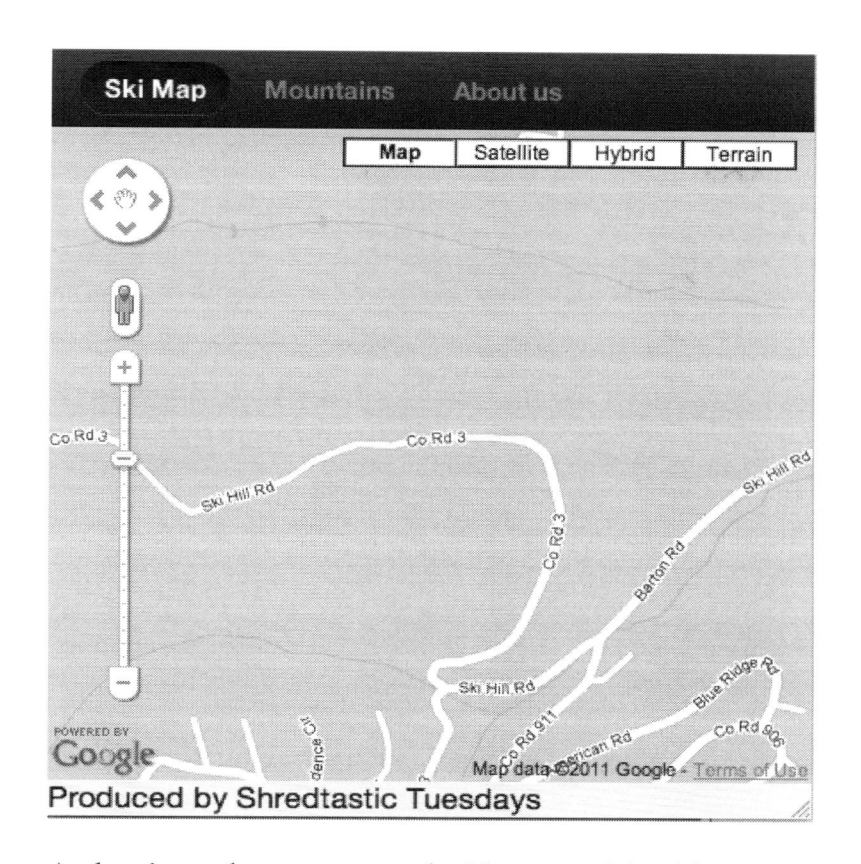

And we've only gotten started with some of the things you can do with Google Maps and Sencha Touch. Let's try some more.

Adding a Marker for Your Business Address

If you're a smart business person, you'll want to point out where your business is located. A pinpoint on a map will make it as clear as possible. Sencha Touch allows you to communicate with Google maps, meaning you can place a pin at any location. To do this, you just need to add a few extra lines of code to your app. Some of it's a bit confusing at first, but don't worry. I'll step you through the process.

```
map.update(
{
    latitude: 39.51,
    longitude: -106.06
}
);

var position = new google.maps.LatLng(39.51, -106.06);

var marker = new google.maps.Marker({
    map: map.map,
    position: position
});
```

Right after your map.update() function, you have to define the place where you want to put your pin. For the sake of this example, I'm going to use the exact same coordinates, 39.51 and -106.06.

What is this google.maps.LatLng() thing? I thought we were using Sencha Touch. Yes we are, but we also included Google's own APIs when we modified our senchademo HTML file to allow us to use maps.

The position variable is basically a condensed set of latitude and longitude coordinates. We have to feed it into our marker because it's the only kind of coordinate information the marker understands. You should also take note of the "map" attribute. We need to tell Google which map to place our maker on. In this case, it's the map we just created.

When you add this slight modification, you should see a nice red marker in the middle of your map.

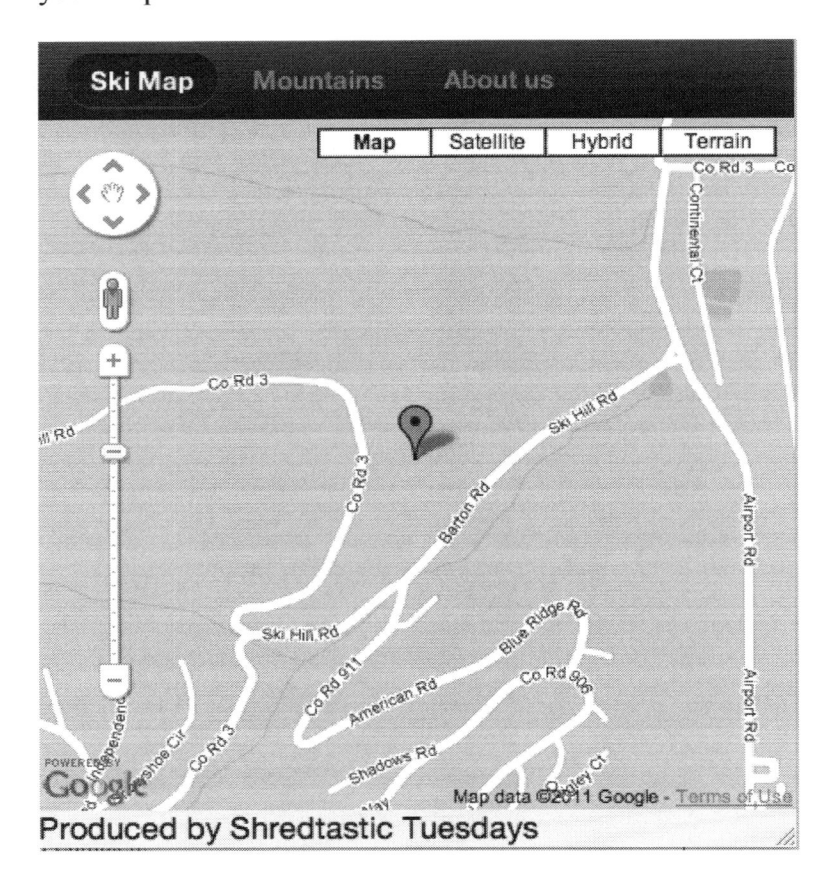

You'll probably want something much more accurate. That's why there are services like batchgeo. You can get your latitude and longitude coordinates by simply typing in your business address. Check it out.

http://www.batchgeo.com/lookup

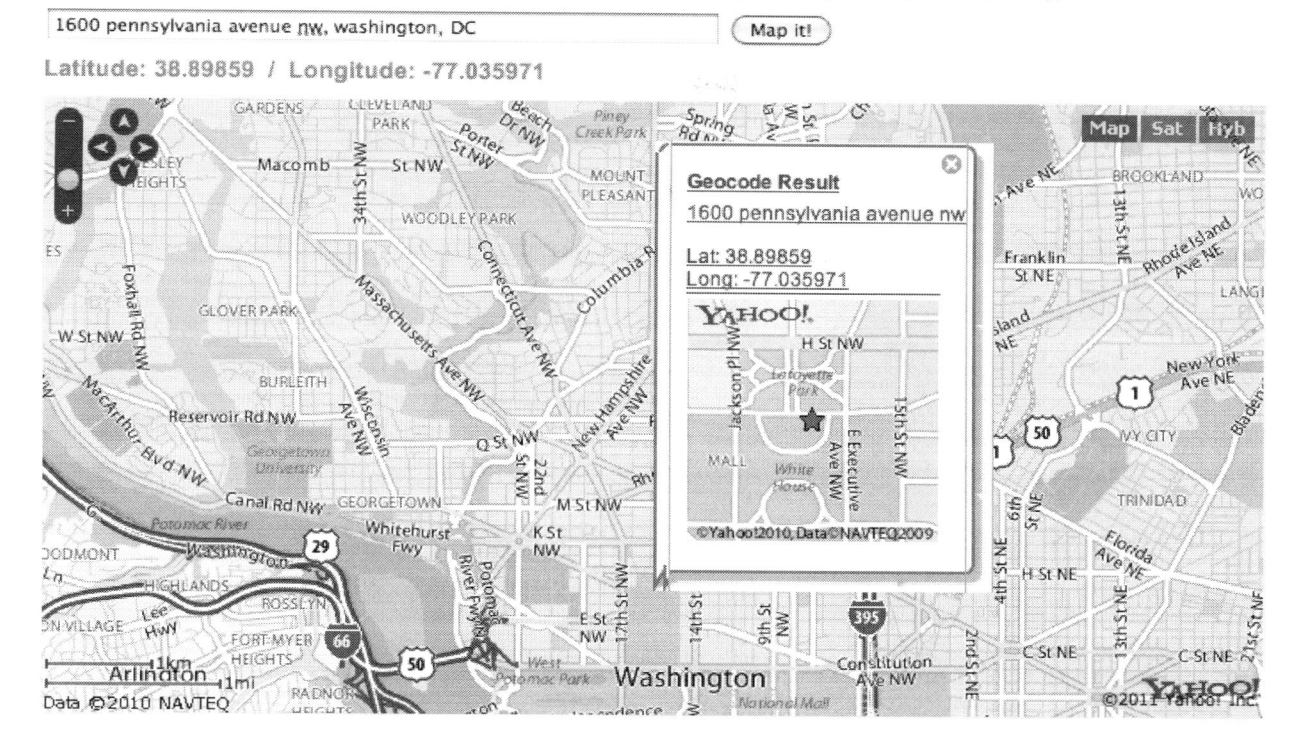

Now we simply plug those coordinates into our app, and we've got a map of the White House.

There isn't much skiing here, but you get the point. It isn't that difficult to show your users where you are located on a map.

Just in case you missed something in the ordering of our code, here is everything you need to make our app so far:

```
 1  Ext.setup({
 2      tabletStartupScreen: 'tablet_startup.png',
 3      phoneStartupScreen: 'phone_startup.png',
 4      icon: 'icon.png',
 5      glossOnIcon: false,
 6      onReady: function(){
 7
 8          var map = new Ext.Map({
 9              title: 'Ski Map',
10              mapOptions: {
11                  zoom: 15
12              }
13          });
14
15
16          new Ext.TabPanel({
17              fullscreen: true,
18              ui: 'dark',
19              sortable: true,
20              cardSwitchAnimation: 'flip',
21              items: [map,
22              {
23                  title: 'Mountains',
24                  html: 'Mountains to ride',
25                  style: "background-color: #f00"
26              },
27              {
28                  title: 'About us',
29                  html: 'Read more about us',
30                  style: "background-color: #0f0"
31              }
32              ],
33              dockedItems: [
34              {
35                  dock: 'bottom',
36                  html: 'Produced by Shredtastic Tuesdays'
37              }
38              ]
39          });
40
41          map.update(
42          {
43              latitude: 38.89859,
44              longitude: -77.035971
45          }
46          );
47
48          var position = new google.maps.LatLng(38.89859, -77.035971);
49
50          var marker = new google.maps.Marker({
51              map: map.map,
52              position: position
53          });
54
55      }
56  });
```

It's pretty amazing if you think about it. In just 56 lines of code, we've built a user interface with interchangeable panels and a map that places a pin at a starting location of our choosing. Life sure is nice when you don't have to reinvent the wheel. Javascript libraries like Sencha Touch and the Google Maps API make it all possible.

Sencha Touch Does Video Too.

There's no better way to market your business than with a video, and Sencha Touch makes it really easy. You can add video with a bunch of different techniques, but we're going to pick the most elegant one. To add a video to our app, we will first create an Ext.Video() object, and then we'll embed it into the items: [ ] array of our TabPanel. All of this should sound very familiar because it's exactly what we did with our Google map.

Before you begin, make sure you have a video file on your server somewhere. Unfortunately, you can't just link to a YouTube or Vimeo file using this method. If you want, you can try the link I've provided for my video.

Here's the code you'll use to create your video object:

```
var snowvideo = new Ext.Video({
    title: 'Video',
    x: 0,
    y: 0,
    width: 200,
    height: 309,
    url: "http://www.shredtastictuesdays.com/senchatouch/dubattempt.mov"
});

new Ext.TabPanel({
    fullscreen: true,
```

I included a little bit of the TabPanel code below, just to remind you that you need to create your video and your map before you embed them into the items: [ ] array. The rest is fairly self-explanatory. You give Sencha a starting x location, a y location, a title, a width, and a height. After that, Sencha does the rest.

As a cautionary note, make sure your width and height are proportional to the actual size of your video. There are a few ways to check this. If you know your video's height and width, you can use Photoshop to do the calculation. Just create a new image with the same dimensions as your video, and go to the image->image size menu item. Once you're there, simply type in your video's smaller width, and you should get the corresponding height.

We just need to do one more thing to put the video in our app. We need to include it in our items: [ ] array, right after the map.

```
new Ext.TabPanel({
    fullscreen: true,
    ui: 'dark',
    sortable: true,
    cardSwitchAnimation: 'flip',
    items: [map,snowvideo,
    {
        title: 'About us',
        html: 'Read more about us',
        style: "background-color: #0f0"
    }
```

Remember, you don't need any fancy curly braces around your map or video objects. Just type in the name of the object and place a comma after it.

There's a reason we're writing our code this way. Simply put, it's more compact and elegant. We could have created the map or the video inside of the items array, but then we'd start to get curly braces inside of more curly braces inside or more curly braces. Our code would have been too complicated to read. And when it's too complicated to read, it's even more difficult to edit.

Sure, it takes a little more time to declare everything up top, but you'll save a ton of time in the long run because you won't have to find the missing curly brace that's messing up your code. Plus, I don't know about you, but I get a warm and fuzzy feeling inside when my creations are elegant like this. You will too.

Produced by Shredtastic Tuesdays

What About YouTube and Vimeo content?

To get YouTube and Vimeo content, we just have to create a container for some HTML. And what container works better than a standard panel? We'll simply change our snowvideo's type to "Ext.Panel()," and then we'll add an extra field for the HTML embed code you get from YouTube or Vimeo. It looks like this:

```
14
15        var snowvideo = new Ext.Panel({
16            title: 'Video',
17            html: '<iframe src="http://player.vimeo.com/video/14235244?portrait=0" width="300" height="169" frameborder="0"></iframe><p><a
   href="http://vimeo.com/14235244">Double Rodeo Attempt</a> from <a href="http://vimeo.com/user3607575">Ted Bendixson</a> on <a href="http://
   vimeo.com">Vimeo</a>.</p>'
18        });
19
```

Once you've entered this bit of code, you don't need to change anything else in your Sencha Touch program. That's the beauty of what we computer programmers call "modularization." When your code is separated into self-sufficient functions, components, and modules, a single change in one place can have a huge impact on your entire program.

Now when we refresh our Sencha Touch app, we get a screen with a standard Vimeo video.

Double Rodeo Attempt from Ted Bendixson on Vimeo.

Cool. We've got a map, and we've got some videos. What else can we do with Sencha Touch? Here's an idea. Let's add a swiping video carousel with a few more videos so our users don't get bored with the one we've got. You'll be surprised that it doesn't take much more code than what we've already written.

Sencha Touch has an component called Ext.Carousel(). It functions just like a panel. It has an items: [ ] array for our videos, so we'll be using it in a very similar fashion.

At this point, we're starting to place panels inside of panels inside of more panels, so it can get a bit confusing if you aren't following closely. If you like, you can imagine our app as a kind of tree or hierarchy. Here's what it would look like.

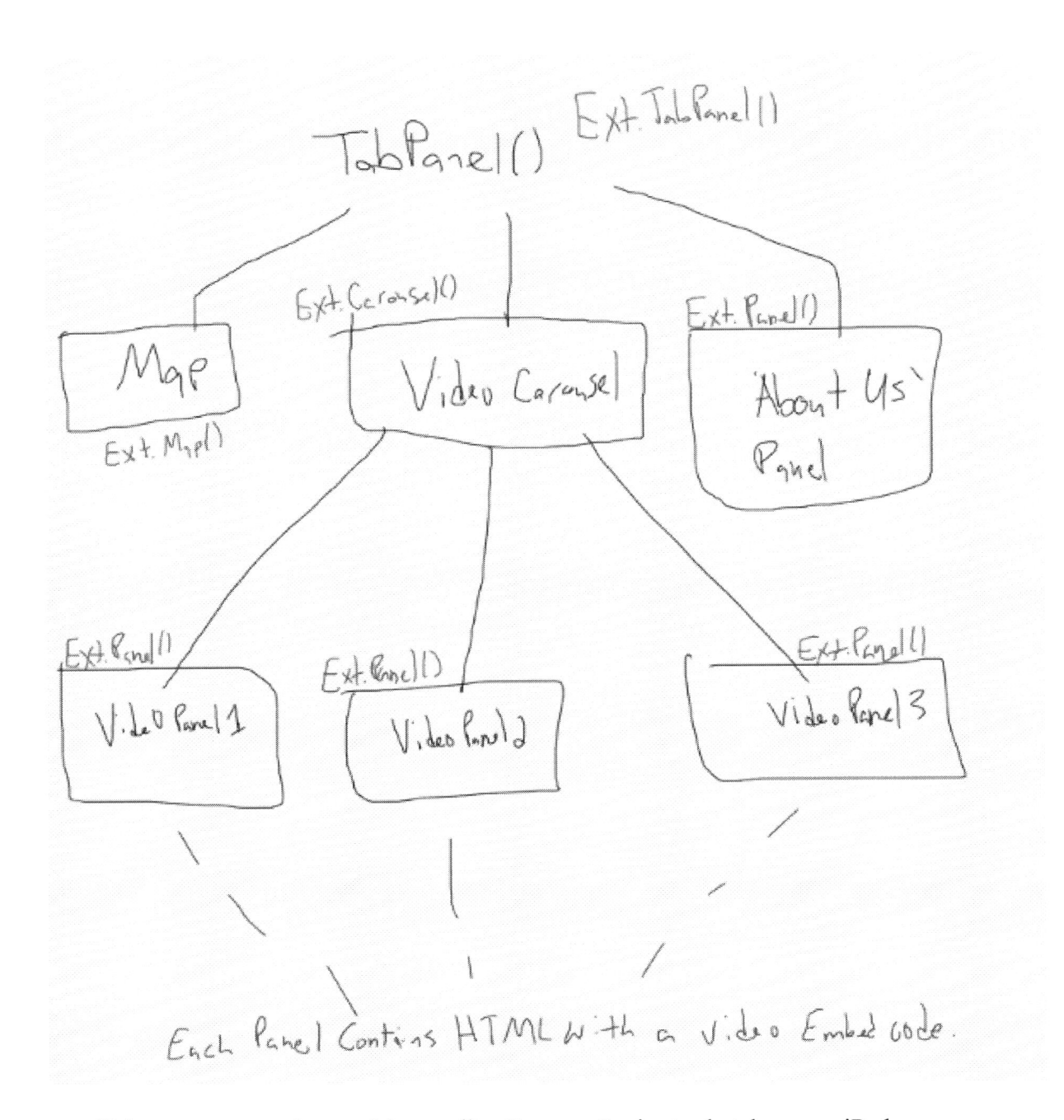

You'll have to excuse the graphics quality. It was a 5 minute sketch on my iPad. Everything is organized in a hierarchy, with our TabPanel() on the top. We'll embed our video carousel in the TabPanel's items[ ] array, and then we'll embed each individual video panel in our video carousel's items[ ] array.

To start, we need to create each individual video panel. So far as the ordering is concerned, we need to do this before we create the video carousel and the TabPanel. If you simply add the next two videos after the first, you'll be in the clear. We already have the ordering from our previous example. We're just building on top of it.

```
15        var snowvideo = new Ext.Panel({
16            title: 'Double Rodeo Attempt',
17            html: '<iframe src="http://player.vimeo.com/video/14235244?portrait=0" width="300" height="169" frameborder="0"></iframe><p><a
.  href="http://vimeo.com/14235244">Double Rodeo Attempt</a> from <a href="http://vimeo.com/user3607575">Ted Bendixson</a> on <a href="http://
.  vimeo.com">Vimeo</a>.</p>'
18        });
19
20        var snowvideo2 = new Ext.Panel({
21            title: 'Elance lets me do this',
22            html: '<iframe src="http://player.vimeo.com/video/14711284?portrait=0" width="300" height="169" frameborder="0"></iframe><p><a
.  href="http://vimeo.com/14711284">Elance Promo</a> from <a href="http://vimeo.com/user3607575">Ted Bendixson</a> on <a href="http://
.  vimeo.com">Vimeo</a>.</p>'
23        });
24
25        var snowvideo3 = new Ext.Panel({
26            title: 'Cardrona NZ Teaser',
27            html: '<iframe src="http://player.vimeo.com/video/13540464?portrait=0" width="300" height="169" frameborder="0"></iframe><p><a
.  href="http://vimeo.com/13540464">Cardrona Teaser Ted Bendixson</a> from <a href="http://vimeo.com/user3607575">Ted Bendixson</a> on <a
.  href="http://vimeo.com">Vimeo</a>.</p>'
28        });
29
```

This should be really easy, especially if you got the previous example to work. I simply copied and pasted my code three times, and then I changed the names of each video panel. After that, I found some different Vimeo embed codes for my other videos, and it all came together in a snap.

The next item we want to create is our video carousel. You can place this code right after your last video.

```
30        var videocarousel = new Ext.Carousel({
31            title: 'Video',
32            items: [snowvideo,snowvideo2,snowvideo3]
33        });
34
```

Are you starting to see how elegant this process can be? We've already done the heavy lifting with the video code above. Now we simply need to wrap it up in a carousel user interface. Remember to keep the title the same. It's the title that will appear in your TabPanel at the top.

Oh, and speaking of TabPanels, we need to make a slight modification to the items[ ] array as a last touch. We'll simply replace "snowvideo" with "videocarousel."

```
new Ext.TabPanel({
    fullscreen: true,
    ui: 'dark',
    sortable: true,
    cardSwitchAnimation: 'flip',
    items: [map,videocarousel,
    {
```

That ought to do it. When you refresh your app and upload your code, you should see three dots at the bottom. Those dots are there to remind your users that they can swipe left or right to reveal more videos. It's so reminiscent of iOS that, if they own an iPhone or any other iOS device, they'll know exactly what to do.

Check out these rad screenshots:

Ski Map   Video   About us

Double Rodeo Attempt from Ted Bendixson on Vimeo.

Produced by Shredtastic Tuesdays

Ski Map   Video   About us

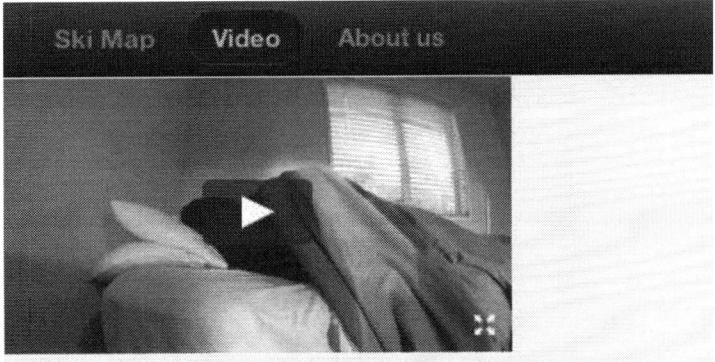

Elance Promo from Ted Bendixson on Vimeo.

Produced by Shredtastic Tuesdays

Well that about covers everything you can do with the user interface elements Sencha Touch offers. There are many other UI widgets you can use in your app, and you can find them by reading through Sencha's API documentation. Just remember to have a plan of attack beforehand, and always keep your functions and components separate. It won't be long before you're placing panels inside of panels and keeping your code nice and elegant.

Buttons and Event Handlers in Sencha Touch. Let the Fun Begin!

Do you remember that temperature conversion app we created with iUI? Well, in this exercise, we're going to build it again -this time with Sencha Touch. Why is this important? Because if you want to build in more complicated functionality, the "brains" of your program, you'll need to know how to trigger other Javascript functions.

I'm going to offer this caveat to those who do not wish to go too deep into this rabbit hole. We're going to start writing some much more advanced code. I will step you through it, as always, but I'm warning you that it won't be 100% intuitive unless you have a little programming experience from other languages. If you have ever done anything in C++, however, this ought to be right up your alley.

In any case, you don't need to know the following in order to build a simple business app. I'm merely showing you this so you can see how you might program in some more complex functionality.

So let's get started. First off, we need to create a panel to display our temperature conversion app. We'll then place this panel inside of our TabPanel, replacing our "About Us" section. At this point, it should be pretty self-explanatory because we've already done it twice. Just declare a new Ext.Panel() and stick it in the TabPanel's items[ ] array.

In this case, I'm going to call my panel tempConverterPanel. As our program gets more complicated, we need to start using variable names that properly describe the elements of our program. This name tells me exactly what I'm working with.

```
var tempConverterPanel = new Ext.Panel({
    title: 'Temperature Converter',
    items: [
    {
        html: 'Temperature in Fahrenheit',
    },fTempBox,
    {
        html: 'Temperature in Cesius',
    },cTempBox,convertBtn
    ]
});
```

Our tempConverterPanel contains a few items that will make up the user interface of our temperature converter. The first element is a simple html title for the first input box. After that, we've got the actual text field for the user's input, which we've called "fTempBox."

fTempBox does exactly what it says it does. It is a box that contains the temperature in Fahrenheit. After that, we've got another bit of html to serve as the title for the second box. That's the temperature in Celsius.

And then there's the last item. The convertBtn is a button just like the one we used in the first Sencha Touch tutorial. When you press it, our app will convert the temperature and display it in the cTempBox.

Have a look at these two variables. These are our text boxes, fTempBox and cTempBox.

```
var fTempBox = new Ext.form.TextArea({
    value: 32
});

var cTempBox = new Ext.form.TextArea({
    value: 0
});
```

Sencha Touch provides a handy TextArea widget that functions just like any other TextArea. A user can provide some input, and then you can pull that input out of the text box and do whatever you like with your program.

In this example, we've simply created the two text boxes and setup some default values. To start, we know that 32 degrees Fahrenheit is equal to 0 degrees Celsius.

We've also got our convertBtn. Have a look:

```
var convertBtn = new Ext.Button({
    text: 'Covert Temperature',
    handler: convert
})
```

Like I said, this button is the same as the buttons we've been using earlier, except there's one difference. This time, we've attached a "handler" to our button. What is a handler? In very basic terms, it's a separate function that we run when the button is clicked. Sencha Touch doesn't use "onClick =" notation. You need to define a handler function and then use the "handler: " attribute to attach that function.

So where is our convert function? I've been saving it for you up until now. This is where our program starts to get interesting. I'm just going to display the entire function, and then we'll step through it slowly.

```
//Convert is a function that converts fahrenheit to celsius
var convert = function(){

    //ftemp is our beginning temperature in fahrenheit
    //ftempBox.getValue() takes the number from the text box
    var fTemp = fTempBox.getValue();

    //cTemp will be our temperature in celsius
    var cTemp = fTemp - 32;

    //After this final step, we get a temperature in celsius
    cTemp = cTemp/1.8;

    //We can use cTempBox.setValue() to put our celsius temperature
    //into the text area.
    cTempBox.setValue(cTemp);
};
```

First of all, have a look at these wonderful green comments! When you type "//" at the beginning of any line, it counts as a javascript comment. We didn't use them before because our programs were fairly simple. But now that we're getting into some advanced techniques, we'll need to use them to remember what everything stands for and how it works.

You should get a fairly decent idea of how our program does its job by simply reading the comments. That's the point of having comments. If you write some function, and you think you might not be able to remember what it does later on, you should probably write a bunch of comments inside of it to jog your memory. Nothing is worse than returning to a project months later only to find yourself completely lost in your own code. Don't let it happen!

Stepping Through Our Convert Function

The first line of this function is already kind of weird. Why are we declaring our function as a variable? I thought the two were supposed to be separate. Well, they usually are, but not when you are attaching a handler to a button. When your function is declared as a variable, Sencha Touch allows you to set it as the handler for any button. That's just how it works.

To start, we're creating a number that we call "fTemp." fTemp, just like all of our variables, does what we say it does. It is the temperature in Fahrenheit.
But to get that number, we need to pull it from our text area. How do we do that? We do it by calling the "fTempBox.getValue()" function. getValue() is a function attached to our text box. Without going into too much detail about object-oriented programming, I can tell you that fTempBox is kind of like a big mashup of functions and information. When

we use a "." after it, we can use some of the functions attached to it. In this case, we're using the one that takes the temperature from the box.

If you really want to get into advanced Javascript programming, I would highly recommend reading a few books on object oriented programming in Javascript. This will give you a better mindset to understand weird things like why our fTempBox has functions attached to it. For the time being, I can only tell you that we need to use fTempBox.getValue() to get the text from the box.

In the next line, we create another variable, "cTemp." This is the number that will represent our temperature in Celsius. If you remember all the way back to our earlier example, we first subtracted 32 from our Fahrenheit temperature, and then we divided it by 1.8 to get our temperature in Celsius.

Some of you might be confused when we use lines of code like "cTemp = cTemp/1.8" How can something like this work? Well, it isn't the same as what you learned in math. cTemp isn't actually equal to itself divided by 1.8. The " = " operator, instead, *assigns a value* to cTemp. In this case, it's assigning the value of itself divided by 1.8.

It's kind of like saying this:

"The *new* cTemp is equal to the *current* cTemp divided by 1.8"

You'll notice that we did the same thing when we subtracted 32 from fTemp. Just like freedom isn't free, equals isn't exactly equals either. It functions more like an *assignment* operator.

Now we're at the last line of code. cTempBox also has a few functions up its sleeves. You can call the setValue() function to put some information into the box. In this case, the information we're putting in the box is our temperature in Celsius (cTemp).

What remains to be done? At this point, we simply need to place our tempConverterPanel into the tabPanel. To do that, we'll add it at the end of our items: [ ] array.

```
new Ext.TabPanel({
    fullscreen: true,
    ui: 'dark',
    sortable: true,
    cardSwitchAnimation: 'flip',
    items: [map,videocarousel,tempConverterPanel],
    dockedItems: [
    {
        dock: 'bottom',
        html: 'Produced by Shredtastic Tuesdays'
    }
    ]
});
```

And here's our final result:

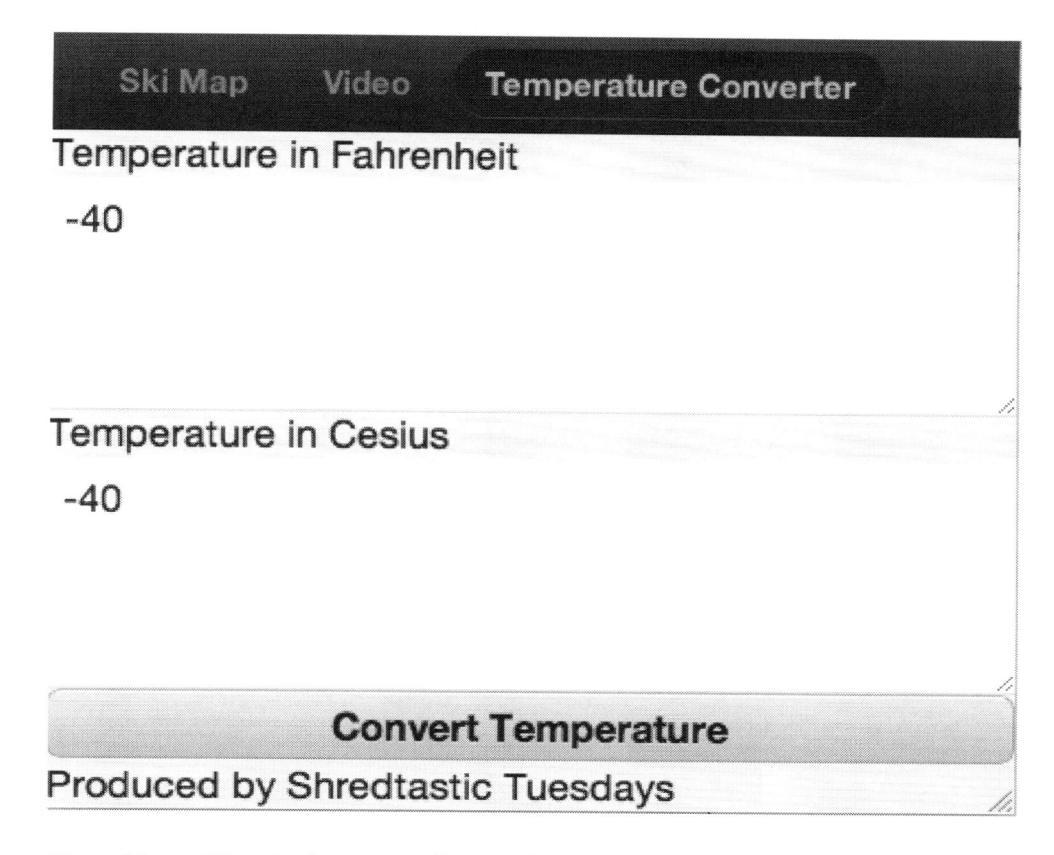

Two things. Yes, it does actually get that cold in the mountains, and yes, -40 in Fahrenheit is equal to -40 in Celsius. Look it up. I'm using it as a test case to make sure our program is doing all of the calculations correctly.

Our Temperature Converter App At a Glance

Here's another reference for you. It's all of the code we're using to build the temperature converter app. Remember that the order in which you declare the variables is important. You can't use something unless it's already defined. Our app is setup so that our functions use the things we've defined beforehand. Here's the proper ordering.

```
var fTempBox = new Ext.form.TextArea({
    value: 32
});

var cTempBox = new Ext.form.TextArea({
    value: 0
});

//Convert is a function that converts fahrenheit to celsius
var convert = function(){

    //ftemp is our beginning temperature in fahrenheit
    //ftempBox.getValue() takes the number from the text box
    var fTemp = fTempBox.getValue();

    //cTemp will be our temperature in celsius
    var cTemp = fTemp - 32;

    //After this final step, we get a temperature in celsius
    cTemp = cTemp/1.8;

    //We can use cTempBox.setValue() to put our celsius temperature
    //into the text area.
    cTempBox.setValue(cTemp);
};

var convertBtn = new Ext.Button({
    text: 'Convert Temperature',
    handler: convert
})
  var tempConverterPanel = new Ext.Panel({
      title: 'Temperature Converter',
      items: [
      {
          html: 'Temperature in Fahrenheit',
      },fTempBox,
      {
          html: 'Temperature in Cesius',
      },cTempBox,convertBtn
      ]
  });

  new Ext.TabPanel({
      fullscreen: true,
      ui: 'dark',
      sortable: true,
      cardSwitchAnimation: 'flip',
      items: [map,videocarousel,tempConverterPanel],
      dockedItems: [
      {
          dock: 'bottom',
          html: 'Produced by Shredtastic Tuesdays'
      }
      ]
  });
```

Typically, the *last* thing in your Ext.setup() function should be your main panel. That's because everything goes into your main panel, and if anything hasn't been defined yet, you'll get a bunch of Javascript errors. I like to think of the ordering as I'm writing the code. If I need to use something in one of my panels, I make some space above the panel for whatever I'm about to use.

Thankfully, you'll know when you've made this mistake. Your Javascript console will tell you that "such and such is not defined." That means you messed up your ordering, and you need to shuffle your code around until you've got it right.

Parting Shots

Well, we've certainly come a long way now. At this point, you should know everything you need to know to build a basic Sencha Touch app that you can then turn into a native app with PhoneGap. To test your Sencha Touch app in PhoneGap, you do exactly the same thing we did with our iUI app. You simply put all of the files into the www folder and allow PhoneGap to work its magic.

In the final section, we'll talk about the day you've been waiting for. At last, you've done all the testing, and you're ready to take your app to the app store. We'll show you all of the final steps you need to take before you can go public with your app. Breath a sigh of relief because the hard part is over. Now you actually get to start making money with this thing.

Chapter 8: How to Deliver Your App to the App Store

At this point, you've tested your app as much as possible in the iPhone simulator. You don't want to add any more functionality, so it's time for the next step. We're going to do some final tests on the actual iOS devices, and then it will be time submit your app for approval. In this section, we'll detail everything you need to do to have a successful launch.

Testing your app on a real device

There are a number of somewhat daunting steps you'll need to go through to test your app on a real iOS device. Apple not only requires you to sign up for the developer program, you have to get a development certificate for each device you plan to use for testing.

So where do you start? The first thing you'll need is your iOS device's unique identifier. You can find this by plugging your device into your Mac and then running Xcode. Right now, I've got my iPad hooked up to my Mac, and Xcode wants to know if I'd like to use it for development purposes.

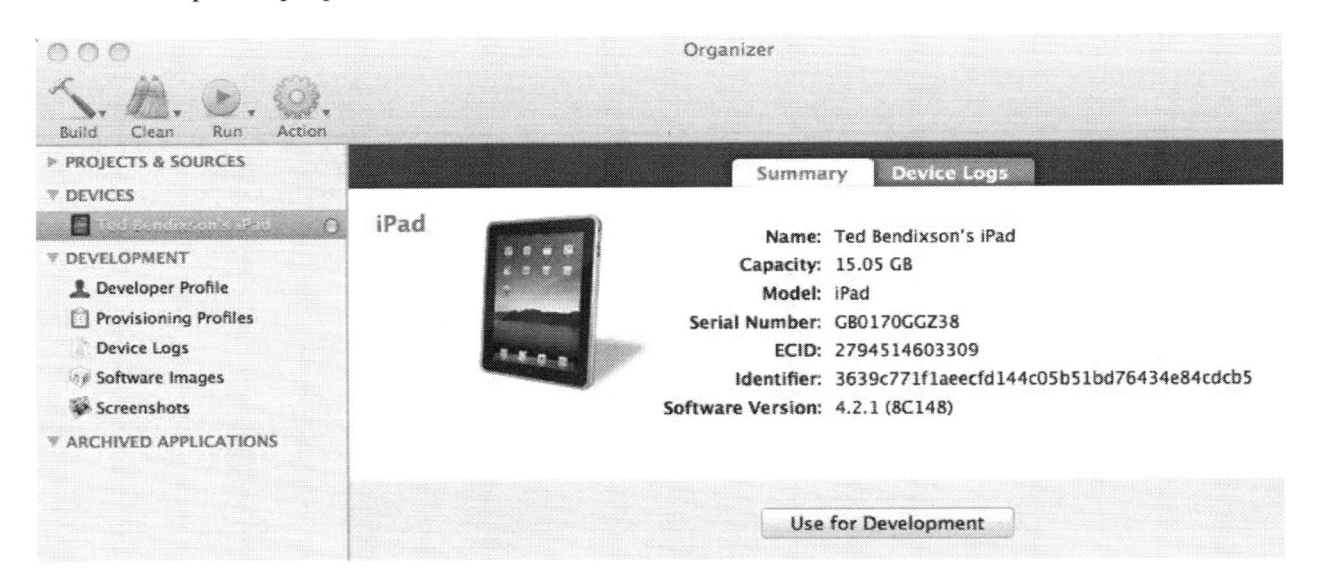

Let's click on "use for Development." After all, it's only the natural thing to do.

At this point, you need to have your Apple ID handy, and you need to have already signed up for the iPhone developer program. If you haven't done so, you'll get an error saying you haven't joined any development teams. Remember, the price for joining Apple's developer program is $99.

We also need to generate a development certificate for our device. This can be done by going to keychain access and requesting a certificate from a certificate authority. Here are a few screenshots to guide you.

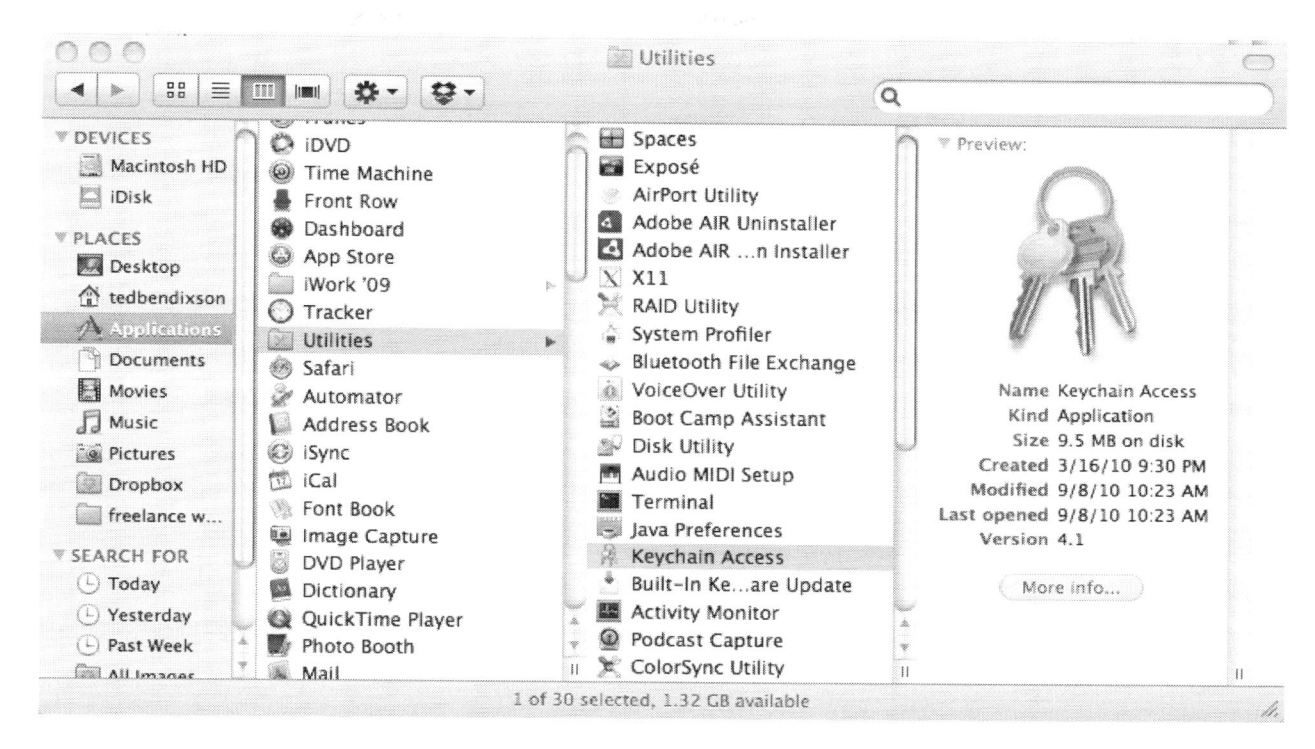

It's right under applications --> utilities.

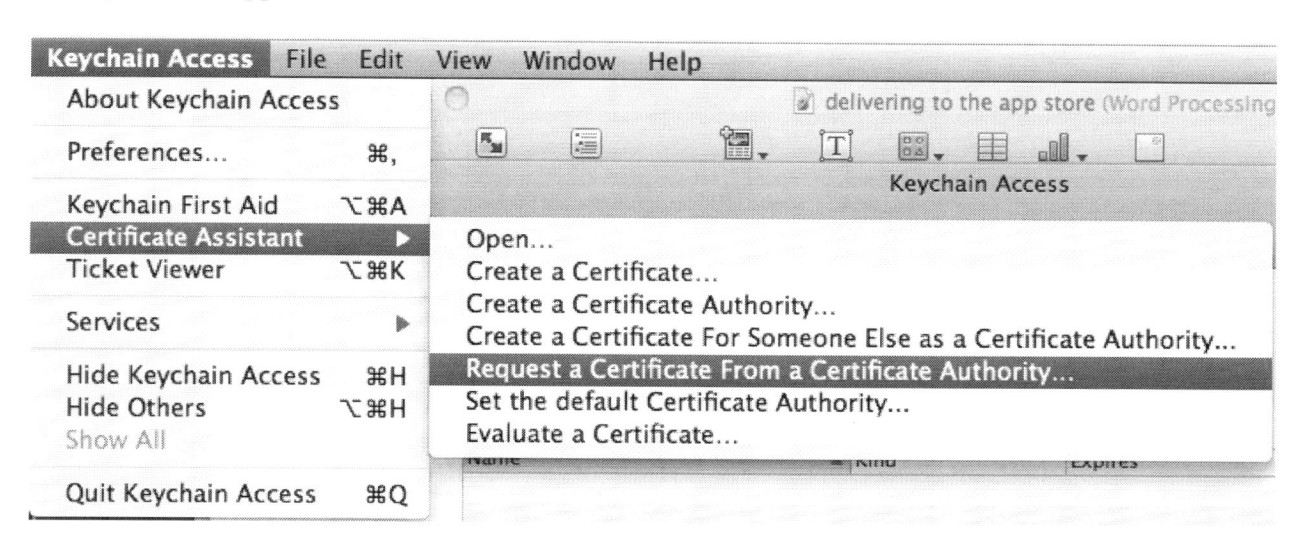

This will launch another setup wizard that will guide you through the certificate request process. Just remember a few things:

1.) Use a key size of 2048 bits
2.) Use the RSA algorithm
3.) Save your certificate request to your disk so you can use it later

With the right certificate, the rest is self-explanatory. You'll be guided through the process, and you'll be able to deploy your app to your device in short order.

A Quick Note on Final Testing.

Be as thorough as you can when testing your app on an actual device. Try to think like your users might think. If a button takes more than a few taps to function properly, most people will put down your app. Better yet, give your device to your friends and see how they use your app. If they don't like it, or they're too bored to keep using it, you need to get back to the drawing board.

The Final Checklist

When you think you've ironed out as many kinks as possible, go iron out some more, and then come back. There are a few more things we need to do before we can get your app to the app store.

To start, we need to check your app's info.plist file in xCode. This file contains some information that will get displayed in the app store, including the icon you plan on using for your app. You can get to info.plist by clicking on "resources" and scrolling down a little bit.

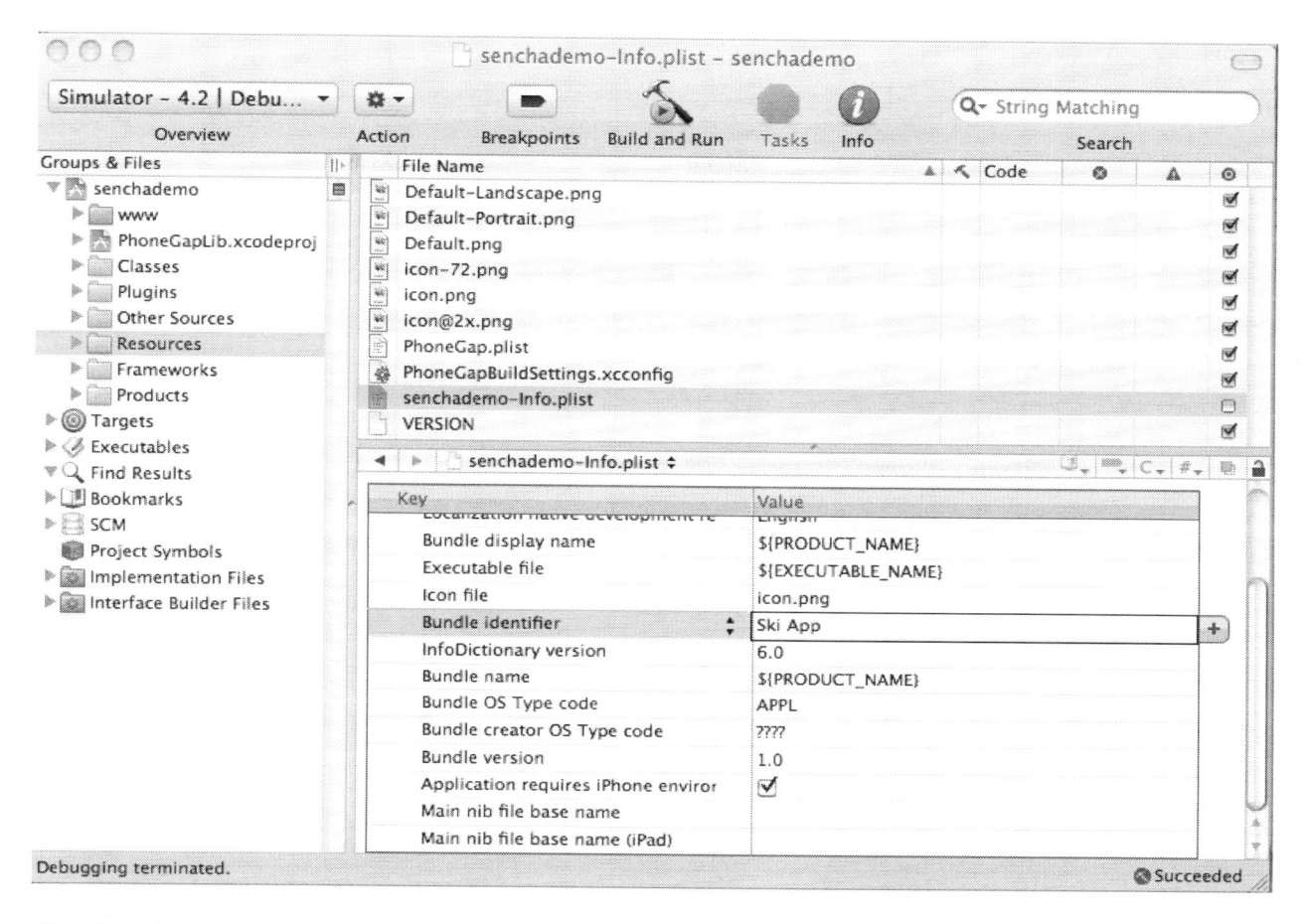

The first thing we want to change is our app's "bundle identifier." We'll simply set it to the name of our app, in this case "ski app." There's also a bundle name field. You can change this if you want your app's name in the app store to be different from the name you've specified in Xcode.

How to Change Your App's Icon from the Default.

As cool as the default PhoneGap icon looks, you'll probably want your own icon for your app. Have another look at your info.plist file. Do you see where it says Icon file and icon.png? Well, icon.png is the current file you're using as your app's icon. If you simply overwrite it with a file of the same name, you change your icon.

Where is icon.png? You can find it by right clicking on the "resources" folder and opening it in the finder.

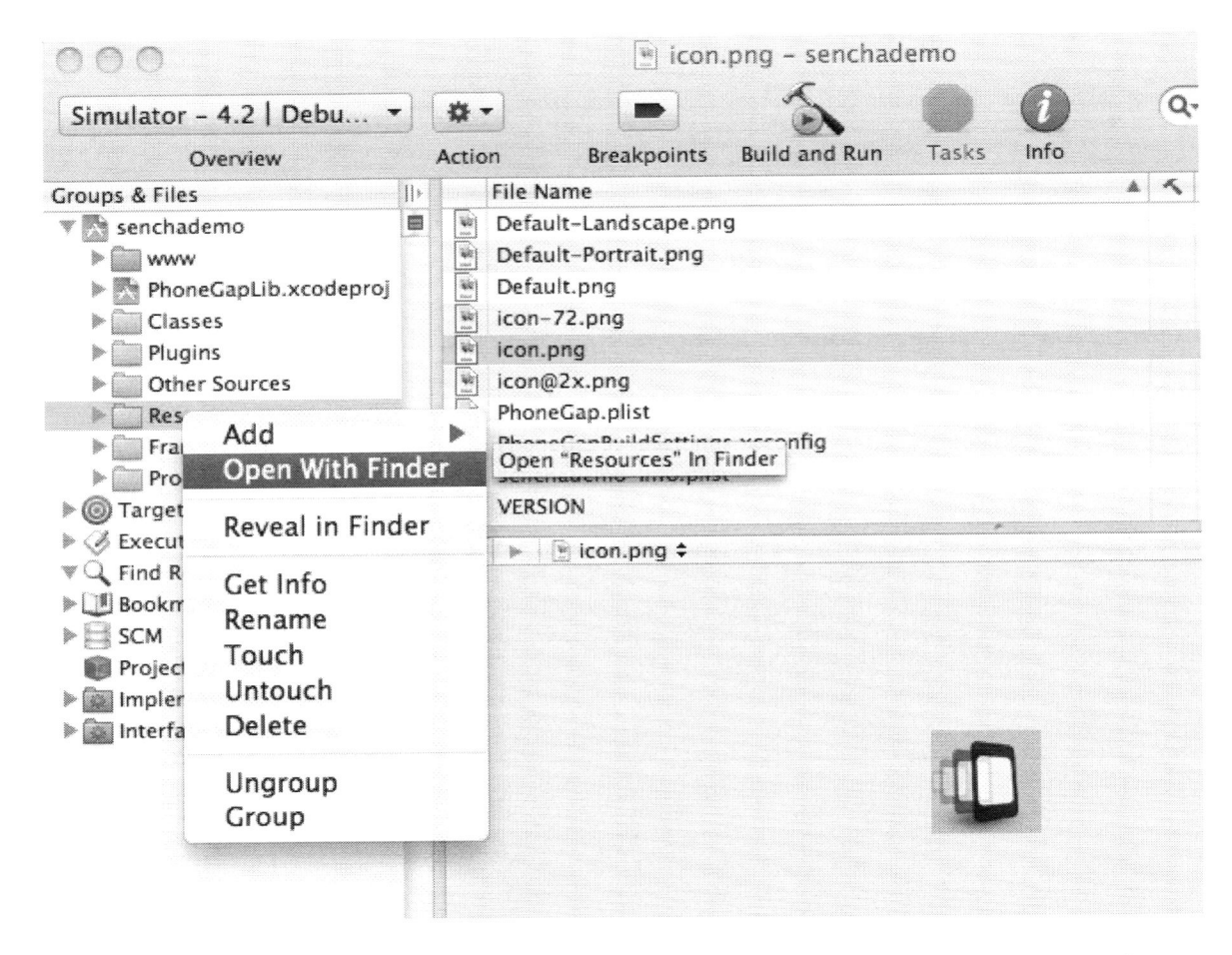

There are a few things you should know about icon.png so you can create your own in Photoshop. The file icon.png is 57 pixels wide by 57 pixels high. There isn't any room for error here. You need to make your icon exactly that size.

Other than that, simply overwriting the file is enough to change your icon. Consider it one more item you've ticked off the list.

Chapter 9: How to Build a Distribution Version of Your App

Up until now, we've been building test versions of our app. With the testing complete, we're ready to build an app that can be distributed to the app store. But before you begin any of this, you need to login to Apple's developer center and get your distribution certificate and provisioning profile. It's another fairly self-explanatory process. Once you login, you'll know what to do.

To make a distribution build, we're going to change a few our app's properties by right clicking on our app under the "targets" menu item in Xcode.

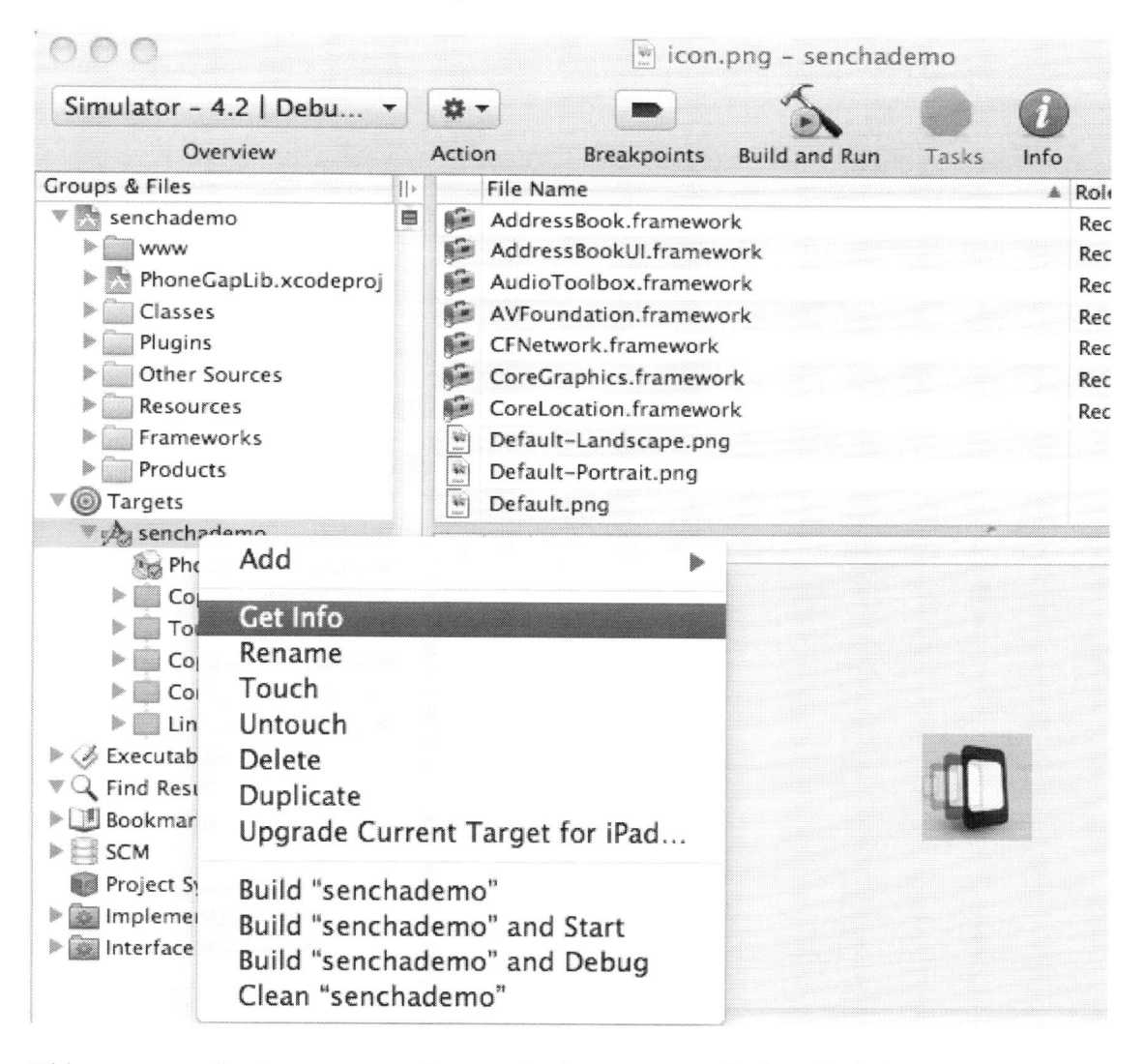

This opens up the target properties panel where we can find a whole host of developer settings. To start, we need to tell Xcode that we want to build a release version of the app. We can find that setting under the "build" tab.

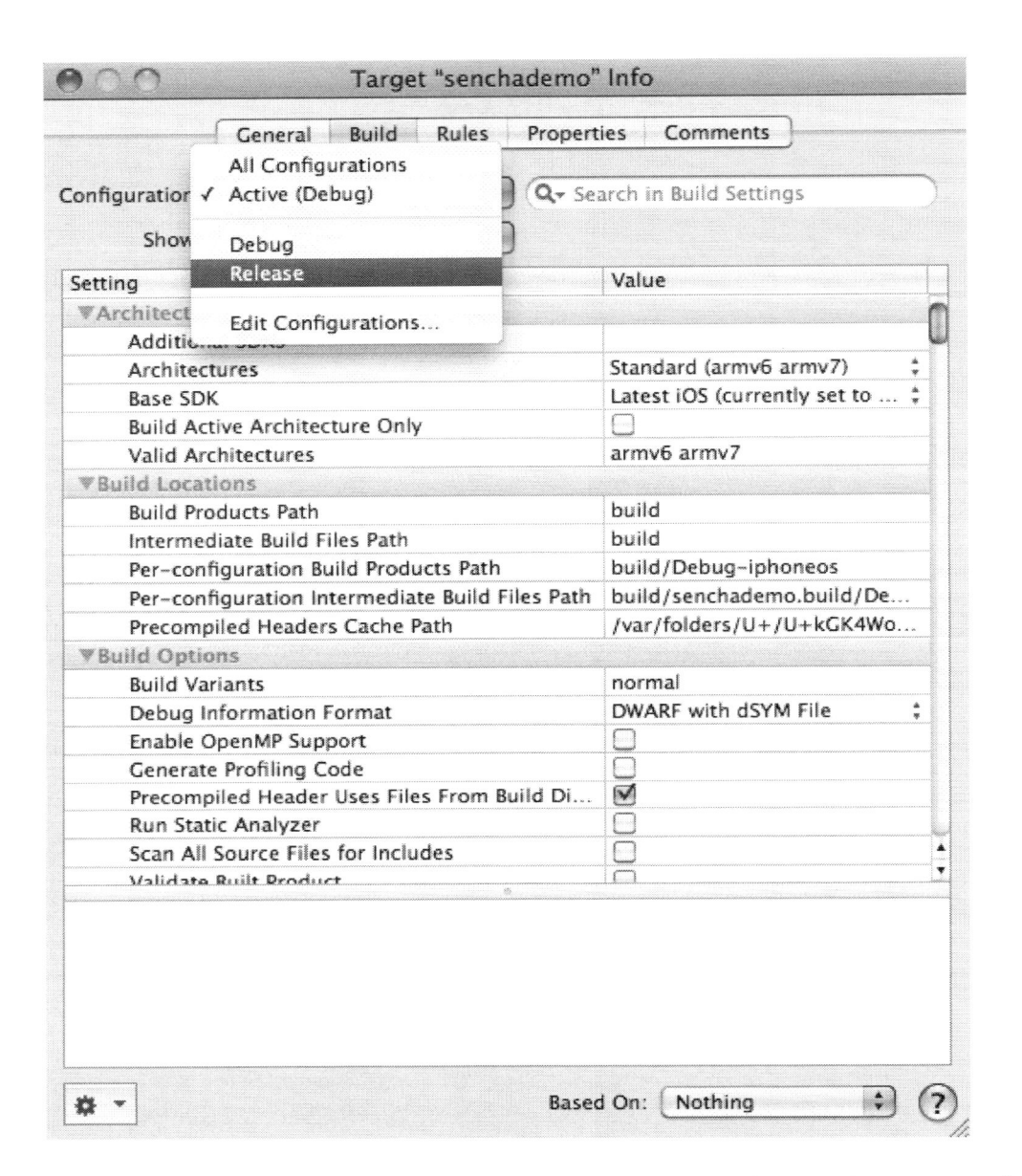

If you scroll down a little further, you'll see the code signing section. We need switch this to iPhone distribution.

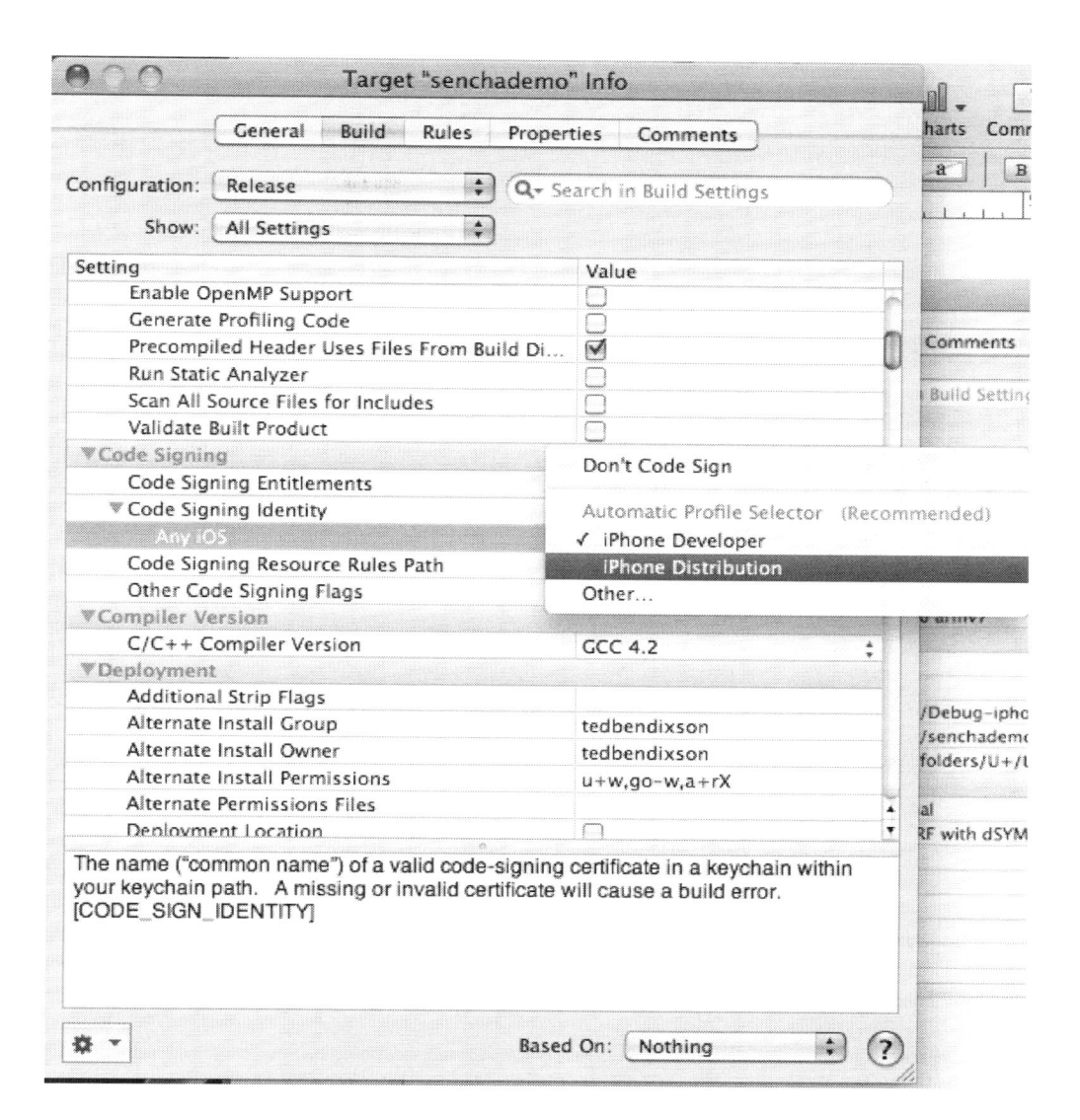

If you have your distribution certificate, you should see your name after you do the switch. If you don't, you'll just see "iPhone Distribution." You need this certificate in order to build a release version of your app, so the sooner the better.

There's one final step before you can build your app's release version. You need to go back to the Xcode toolbar and switch from simulator mode to release mode.

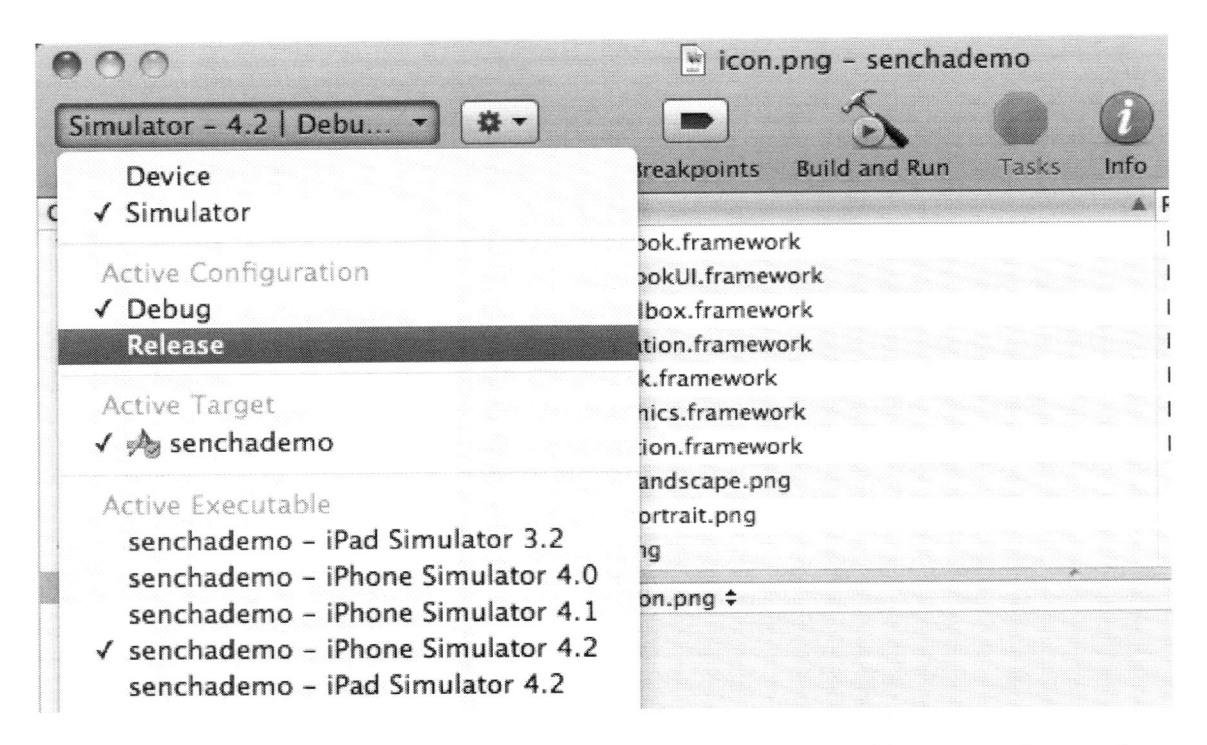

Now click "Build and Run" to make your release version. Once this process is complete, your app will appear in your Products folder.

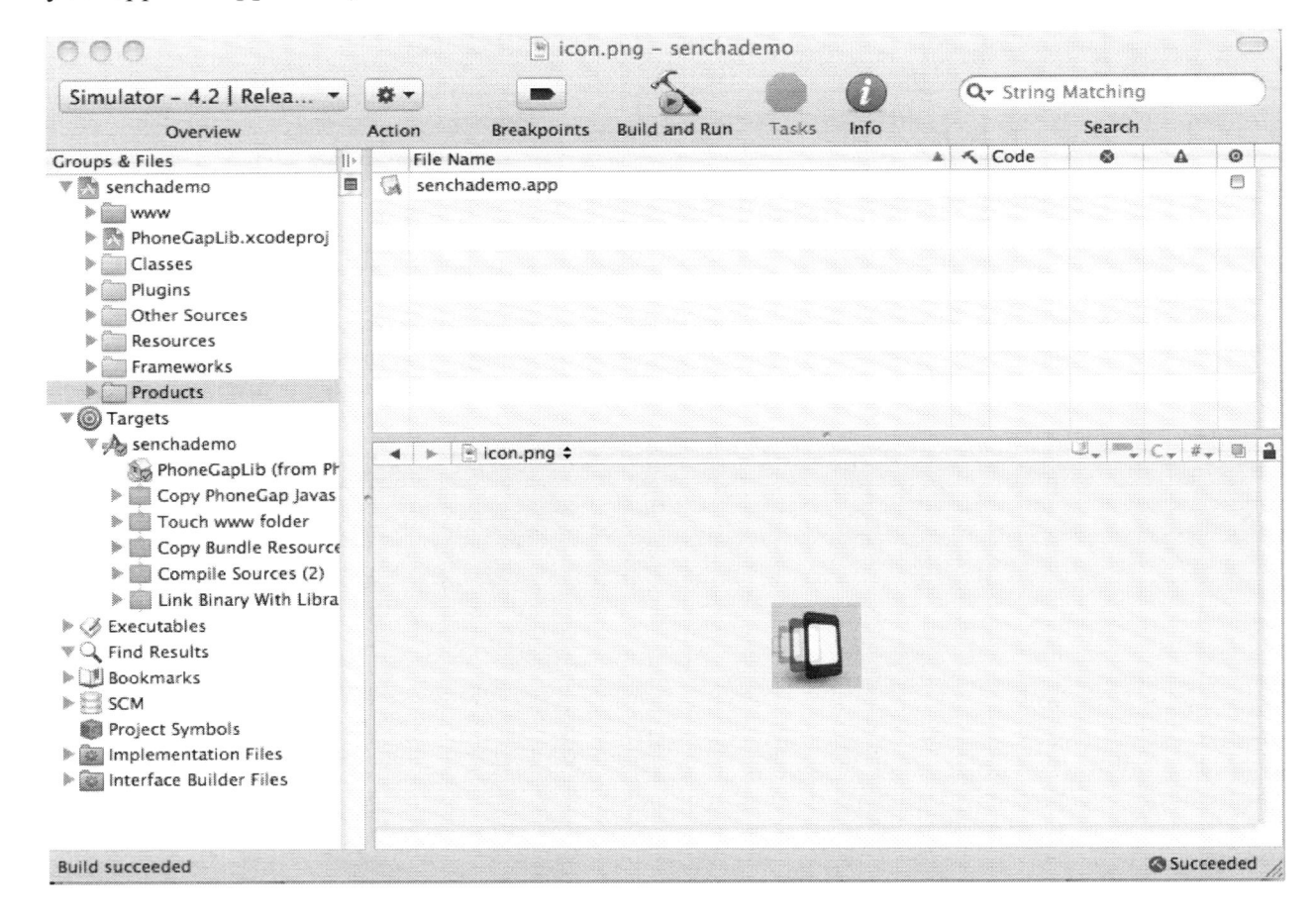

Watch as your app is being built. You need to make sure it gets past the code signing process. If it doesn't, and you submit your app to the app store, it will almost certainly get rejected (which means a few more weeks of waiting). Get it right the first time, and you'll be well on your way.

You can open the products folder by right clicking on it and viewing it in the finder. You'll need to create a zip file of your app in order to submit it to the app store. This is also done by right clicking on your app and choosing compress "appname.app".

Chapter 10: A Brief Look at Marketing

We're running at a quick clip, heading straight for the app store. There's one final stop along the way. Yes, I know, your app is awesome and people will want to use it regardless of your marketing efforts. But I don't care about your excuses, you need to at least prepare *something* to whet your users' appetite.

Get these together:

1.) Four highly appealing screenshots from your app. Consider the following. Which screenshots make you the most curious about your app? Those are the ones you want to pick for the app store.

2.) A 700 character app description. Use bullet points near the top. Each bullet needs to clearly communicate one of your app's features. If you're not the best writer, I highly recommend hiring a copywriter to take care of this.

That's it for now. It wasn't so bad, was it? You'll be glad you did this when your app hits the app store. It's good to have momentum on your launch date, and anything that gives your buyer more information about your product helps.

Off to the App Store!

And away we go. Login to your Apple developer center account, and click on the "iTunes connect" link. From there, click on the "manage applications" link, and then "add new application."

Pretty soon, you'll be asked a dizzying array of questions about you, your app, your company, your copyright info, and much more. Most of it's a breeze, but the following may concern you.

1.) You can pick any SKU that you please as long as it's a series of four numbers.

2.) The "application URL" should be the website you intend to use to market your app.

3.) The "support URL" should link to the email account where you'll be taking support inquiries.

4.) You need to upload your screenshots in the opposite order in which you want to display them on your app's page. You should upload your weakest screenshot first and your best screenshot last.

Once you've reviewed your information and clicked the "submit" button, you have nothing else to do but wait. Apple will let you know via email when your app has been approved. It can take weeks or months. Nobody really knows why or how. Apple will also let you know when there's a problem with your app.

In our experience, the approval process has only gotten smoother with time. It's quite unlikely that you'll end up waiting months. This is a great time to get out there and market your app. You've been so busy building the darn thing that you probably haven't spent much time getting the word out. If you don't have a website, build one. If you don't have a Facebook presence, get started now. Oh, and there's also this thing called actual conversations with real people. Those are pretty cool too.

One Last thing.

Yay! Your app is approved, and it's in the app store. Don't pop open that champagne quite yet. If you remember to do this one thing, you'll dramatically improve your app's visibility on launch day.

On the day of your app's approval, go back to your iTunes connect account and change your app's availability date to today. That way, it will appear at the top of the "new releases" section. Those who are curious will browse, and some of them will download your app. That's much better than what happens when your app appears to be a few days old. You need to get as much out of your launch as you can.

Chapter 11: Where do we go from here?

Now that your app is on the iTunes app store (and you're a gazillionaire from all the sales), why not take your app to the Android market and the Blackberry? With PhoneGap it's a hop, skip, and jump away. You've already built your app. To get onto the other stores, you need only learn the submission and build rules for the different platforms (basically the very end of the build process). It's surprisingly easy.

You can start with our Android + PhoneGap development guide. You'll recognize most of the content from this guide. The Android development guide will show you how to use Sencha Touch to build an app that you can deploy to Android market with PhoneGap. Since you already know Sencha Touch, you can skip a few steps and head straight for Android deployment. Easy.

I hope you've picked up a few things in the process of reading this guide. And if you haven't started building your app, do it now! There's no time like the present, and inspiration is fleeting. Go with your gut. Put in a long hard coding session and get the bulk of your app built in one day. You can do it. And when you do, you'll be part of a growing phenomenon in smartphone development.

Today signals the end of native smartphone development. With PhoneGap, you build one mobile app for every platform. One app for the web, one app for iPhone, one app for Android. One app to be quickly deployed wherever you want. One app to rule them all!

Made in the USA
Lexington, KY
28 July 2011